Materials Forming, Machining and Tribology

Series editor

J. Paulo Davim, Aveiro, Portugal

For further volumes:
http://www.springer.com/series/11181

J. Paulo Davim

Editor

Machining of Titanium Alloys

 Springer

Editor
J. Paulo Davim
Department of Mechanical Engineering
University of Aveiro
Aveiro
Portugal

ISSN 2195-0911   ISSN 2195-092X   (electronic)
ISBN 978-3-662-43901-2   ISBN 978-3-662-43902-9   (eBook)
DOI 10.1007/978-3-662-43902-9
Springer Heidelberg New York Dordrecht London

Library of Congress Control Number: 2014943742

Printed on acid-free paper

Springer is part of Springer Science+Business Media (www.springer.com)

Preface

Currently, titanium alloys are in important group of engineering materials due to their excellent combination of strength and fracture toughness as well as low density and very good corrosion resistance. These materials have received special attention recently due to their wide range of applications in aerospace, aircraft, automotive, chemical, and biomedical industries. However, these expansive materials present poor machinability because of their low thermal conductivity and high chemical reactivity with cutting tool materials.

The purpose of this book is to present a collection of examples illustrating research in machining of titanium alloys. Chapter 1 of the book provides machinability and machining of titanium alloys (a review). Chapter 2 is dedicated to cutting tool materials and tool wear. Chapter 3 described mechanics of titanium machining. Chapter 4 contains information on analysis of physical cutting mechanisms and their effects on the tool wear and chip formation process when machining aeronautical titanium alloys: Ti-6Al-4V and Ti-55531. Chapter 5 is dedicated to green machining of Ti-6Al-4V under minimum quantity lubrication (MQL) condition. Finally, Chap. 6 is dedicated to ultrasonic-assisted machining of titanium.

This book can be used as a research book for final undergraduate engineering course or as a topic on manufacturing engineering at the postgraduate level. Also, this book can serve as a useful reference for academics, researchers, mechanical, manufacturing, industrial and materials engineers, professionals in machining technology and related industries. The interest of scientific in this book is evident for many important centers of the research, laboratories, and universities as well as industry. Therefore, it is hoped this book will inspire and enthuse others to undertake research in machining technology.

The Editor acknowledges Springer for this opportunity and for their enthusiastic and professional support. Finally, I would like to thank all the chapter authors for their availability for this work.

Aveiro, Portugal, May 2014   J. Paulo Davim

Contents

Chapter 1
Machinability and Machining of Titanium Alloys: A Review

Seyed Ali Niknam, Raid Khettabi and Victor Songmene

Abstract This chapter reviews the main difficulties impairing the machinability of titanium alloys. The overview of machinability of titanium alloys is presented with respect to the following performance criteria: cutting tool wear/tool life, cutting forces, chip formation, and surface integrity attributes, mainly surface roughness. Thereafter, the effects of various lubrication and cooling methods in machining titanium alloys is also discussed. Furthermore, a case study on the metallic particle emission when machining Ti-6A1-4V is also presented.

1.1 Introduction

Due to their unique properties, titanium alloys are used in major industries, including the biomedical, aeronautic and automotive industries. Compared to aluminum alloys which are also used in several industries, titanium alloys have remarkable mechanical properties, such as superior strength to weight ratio and exceptional corrosion resistance at elevated temperature. The titanium alloys are known as expensive materials when compared to other metals. The main reasons arise from the extraction process and difficulties encountered in melting and fabrication processes [1]. Despite recent developments and extensive usage of titanium alloys, machining of titanium alloys still remains as a major industrial

S. A. Niknam · V. Songmene (✉)
Department of Mechanical Engineering, École de technologie supérieure (ÉTS),
Montreal, QC H3C-1K3, Canada
e-mail: victor.songmene@etsmtl.ca

S. A. Niknam
e-mail: seyed-ali.niknam.1@ens.etsmtl.ca

R. Khettabi
Department of Mechanical Engineering, Université de Guelma, Guelma, Algeria
e-mail: khettabi@uqtr.ca

J. P. Davim (ed.), *Machining of Titanium Alloys*,
Materials Forming, Machining and Tribology, DOI: 10.1007/978-3-662-43902-9_1,
© Springer-Verlag Berlin Heidelberg 2014

Table 1.1 Some titanium alloys used in industry

Alloys	References
Ti-64	Chen et al. [24], Che-Haron and Jawaid [25], Wang et al. [26], Nurul-Amin et al. [27], Sun et al. [28], Sun and Guo [5], Thomas et al. [29]
Ti-6242S	Ginting and Nouari [30], Ginting and Nouari [31], Che-Haron et al. [46]
Ti-6246	Che-Haron [32]
Ti-834	Sridhar et al. [33], Thomas et al. [29]
Ti-45-2-2	Mantle and Aspinwall [34], Mantle and Aspinwall [35]
Ti-4.5Al-4.5Mn	Zoya and Krishnamurthy [36]
Ti-6-6-2	Kitagawa et al. [37]
TA-48	Nabhani [38]
Ti-6Al-7Nb	Cui et al. [39]

concern since they are classified as difficult-to-cut materials (tool wear, vibrations, low metal removal rates, etc.). Titanium alloys still suffer from a poor machinability compared to other metals owing to several inherent mechanical and material properties. The main reasons of relative low machinability of titanium alloys are due to their low thermal conductivity, high chemical reactivity and low elasticity modulus [2]. They are characterized by a high cutting temperature, a short tool life and a high level of tool vibration [1, 3–5].

To overcome the major difficulties in machinability of titanium alloys, research studies in the past decades have paid special interests into (1) development of new cutting tool materials (2) improvement of the existing tool design (3) conducting extensive machinability studies to investigate the right combination of cutting tool, machine tool and cutting parameters [1, 3, 6–15] and (4) development and implementation of advanced hybrid machining processes e.g. laser-assisted machining (LAM), cryogenic, cooling and lubrication systems and heat treatment conditions [13–23]. There are more than 10 titanium alloys (see. Table 1.1) that are widely used in various industrial sectors [5, 24–39]. Amongst titanium alloys, the Ti-6Al-4V is known as the most highly used alloy (50 % of titanium production) [12], which has been also the subject of the machinability evaluation, experimental and modeling studies in the most of the reported works in the literature [10, 12, 19, 30, 40–42]. Further descriptions of factors governing machinability of titanium alloys and the solutions to improve it are very much recommended and could be considered as the subjects of future works.

1.2 Overview of Machinability of Titanium Alloys

Machinability of a material is usually determined based on criteria such as tool life, tool wear, cutting force, chip formation, cutting temperature, surface integrity and burr size. Titanium alloys are classified into four main groups as follows: (1) α; (2) near α; (3) αβ and (4) β [3]. Interested readers may find more detailed descriptions of titanium alloys and their properties in [1, 43]. In general, machining titanium alloys include milling, turning and drilling operations. However, most of the machining studies for titanium and its alloys have been focused on the turning process. The obtained results of turning are not applicable to the milling process, where interrupted cutting takes place, subjecting the tools to a variety of hostile conditions [44].

During milling operation, the cutting tool is imposed to various failure modes due to extensive loading and unloading effects [45]. This phenomenon adversely affect the tool life, cutting forces, surface quality, dimensional accuracy and economic of machining operations. This implies an appropriate selection of cutting tools, machining parameters and lubrication and coolant conditions during machining titanium alloys. To that end, the correct selection of titanium alloys with respect to operation used must be conducted precisely. In the following section, the main properties impairing the machinability of titanium alloys will be introduced using combination of the following performance criteria: cutting tool wear/tool life, cutting forces, chip formation, part quality mainly surface roughness. Furthermore, the effects of various lubrications/coolant modes on machinability of titanium alloys will be presented.

1.2.1 Cutting Tool Materials and Wear Mechanisms When Machining Titanium Alloys

The performance of a cutting tool is normally assessed in terms of its life on the basis of certain wear criterion, mostly flank wear that largely affects the stability of the cutting edge and consequently the dimensional tolerance of the machined work surfaces [36]. The main tool materials used in machining titanium alloys include: (1) uncoated and coated cemented carbides (WC/Co) (2) polycrystalline diamond (PCD) and (3) polycrystalline baron nitride (PCBN) and cubic-boron-nitride (CBN). Amongst described tools, the uncoated and coated carbides tools are the most frequently used tool materials. To protect the tools against wear, various coating materials such as TiN, TiCN, TiAlN and many others such as Al_2O_3 are widely employed [47]. The appropriate selection of coating materials depend on the workpiece material used, whereas the machining of titanium alloys is a thermal dominant process and a critical temperature of 700 °C can be considered as a tool life criterion [36].

The tendency of titanium alloys to react with the most of cutting tool materials is the main factor hindering the machinability of titanium alloys. To reduce the chemical reaction between the tools and material, Ezugwu et al. [48] performed cutting operations in inert enriches environment and observed better tool life under conventional machining environment. According to [49], uncoated tools reach their maximum process temperature faster than coated tools. On the other side, according to [31, 38, 50], during machining titanium alloys, the tool coating rapidly fails, mainly due to high temperature generation between tool and chip and also coating delamination. This exhibits that according to cutting conditions used, various tool performances can be observed from coated and uncoated tools during machining of titanium alloys. Furthermore, in addition to tool wear, machining dynamics and the surface finish of the work part can be used as indicators for evaluating the cutting performance. The tool flank side is the main subject of tool wear in most of the cases in both coated and uncoated tools [46]. It is generally proposed to use the sharp, positive edge tools with ample clearance as well as stable cutting conditions with well-clamped work parts, in order to secure high performance cutting with minimized vibrations tendencies during machining titanium alloys [51].

As pointed out in [32], the majority of tool failure mechanisms during turning Ti-Al-2Sn-4Zr-6Mo were due to flank wear and excessive chipping on the tool flank edge. The severe chipping and flaking of the cutting edge resulted from high thermo-mechanical and cyclic stresses seem to be the main failure mode when milling titanium alloys with carbide tools [44]. Whereas, to reduce the tool wear in milling operations, it is suggested to use low cutting speed [52]. As observed in [52], the wear progress of the coated tools was slower than that of the cemented carbide tool during high speed milling of titanium alloys. According to [45], cutting speed and depth of cut were identified as the main factors responsible for the failure and fatigue of the coated carbide tools during milling of titanium alloys. This causes by the sudden loading which leads to micro fracture and brittle fracture and consequently propagates until total fracture occurs.

Zareena and Veldhuis [53] used single-crystal diamond tools in ultra-precision machining of commercial pure titanium (CP-Ti) and Ti-6Al-4V alloys and then examined the tool wear mechanisms. To reduce the wear mechanism, a protective barrier made of Perfluoropolyether (PFPE) polymer was explored. As concluded in [53], the cutting temperature and high pressure at the tool-chip interface, built up edge (BUE) formation and chemical interaction between the work part and tool are the main reasons of the tool wear. According to [45], better carbide cutting tool properties were obtained when using ultrafine carbide cutting tool grades in conjunction with advanced sintering processes. Although diamond has a high hardness, low friction coefficient and high thermal conductivity, but since it easily reacts with metallic and ferrous work parts, it is not advised to use it as a coating in most of the machining cases. Therefore, more attention has been paid to employ the PCD tools which offer better surface finish and less wear rates compared to PCBN and coated carbide tools [9, 27, 38, 47, 54].

The CBN tools also tend to show lower wear rates, strong thermal shock resistance, higher hardness and strengths, good thermal conductivity, and superior mechanical properties and wider industrial applications compared to carbide tools [26, 54]. However due to chipping phenomenon, it is suggested to use the CBN tools at lower feed rates, which effectively lower tool wear rate and improve the tool life [36]. On the other side, Ezugwu et al. [55] found that during high speed finish turning of Ti-Al-4V titanium alloy (up to $250 \text{ m} \times \text{min}^{-1}$) when various coolants are supplied, the CBN tools have a shorter tool life than that observed in uncoated carbide tools. The recently developed CBN tools known as binderless CBN (BCBN) tools seem to provide improved tool life, surface finish and lower cutting forces compared to that observed when using CBN tools. However these tools are more expensive than CBN tools, but since the tool life is also improved when using BCBN tools, the cost is not a main issue in large production [9, 26]. Based on the literature, although the use of coated tools seems to have positive influences on tool life, tool wear rate and attainable cutting speed [56], but surface integrity attributes, including surface defects, surface finish and residual stresses are not significantly improved when coated tools are used. Furthermore, as reported in [56, 57] the coated tools generate the residual stresses while the use of uncoated tools led to compressive residual stresses that is more favourable during machining operations.

1.2.2 Part Quality and Surface Finish When Machining Titanium Alloys

The surface integrity attributes of titanium alloys work parts are largely affected during machining operation [32–35, 58–60]. The main surface integrity concerns appear on (a) topography characteristics such as textures, waviness and surface roughness (b) mechanical properties affected such as residual stresses and hardness, and (c) metallurgical states such as micro-structure, phase transformation, grain size and shape, inclusions, etc. An extensive research work reported the surface integrity during machining titanium alloys [4, 5, 23, 25, 31–36, 56–66]. Amongst the above-mentioned surface alterations, the recent works on surface finish and part quality of titanium parts under various machining modes will be presented in this chapter. Generally low surface quality is resulted in titanium alloys, so post processing methods such as laser shock peening and ball burnishing are required [58]. In general, minimizing the surface roughness in machining titanium alloy is considered as a topic of current interest, and it has been received huge amount of interests [63–67]. It has been found that the surface roughness in titanium work parts are widely affected by various phenomenon in machining operations, such as built up edge (BUE) formation [35], tool shape, geometry and tool wear [26, 37, 50, 53, 68], temperature [58, 69], tool coating [31, 44, 46, 70], feed rate [5, 31, 63], cutting speed [5, 25, 31, 36] and depth of cut [71]. These effects mainly appear due to thermal and mechanical cycling, microstructural transformations, and mechanical and thermal deformations during machining processes [72].

Ginting and Nouari [31] studied the surface integrity of titanium machined parts in dry milling operations. The roughness lay, defects, micro hardness and microstructure alterations were investigated. It was found that the uncoated CVD tools produce better surface roughness values compared to those recorded when using coated carbide tools. Moreover, dry machining can be carried out with uncoated carbide tools for titanium alloys, if cutting condition is limited to finish and/or semi-finish operations. For instance, in [63], experimental tests were planned under Taguchi's orthogonal array and cutting tests were conducted using CVD (TiN–TiCN–Al$_2$O$_3$–TiN) coated carbide insert under different cutting conditions. It was found that higher feed rate increases the surface roughness, but the increased cutting speed and depth of cut decreases the surface roughness. According to [36], adequate surface finish can be obtained during machining titanium alloys at the speed range of 185–220 m $\times$ min^{-1}. Regardless of all other cutting parameters used, the surface roughness results are found to be higher in fresh cutting tools as compared to those measured when using slightly used cutting tool [58]. However, in general, it is agreed upon that higher surface roughness is observable for worn tools.

Although several studies reported the experimental evaluation of the factors governing surface integrity attributes in machining titanium alloys, however, it is still believed that limited analytical methods have been formulated to calculate the surface roughness for titanium alloys [58]. This lack could be remedied by developing FEM based simulation models for predicting machining induced white layer formation, micro hardness, residual stress profiles and surface roughness for titanium alloys.

1.2.3 Cutting Forces When Machining Titanium Alloys

The factors governing cutting forces during machining titanium alloys have been widely reported in literature [12, 28, 73–75]. According to [73], the use of cutting tools with dull or improperly ground edges increases cutting forces. As shown in Fig. 1.1, the dynamic cutting forces under different cutting speeds, feed rates and depth of cut were measured during dry turning of Ti-6Al-4V titanium alloy [28]. The amplitude variation of the high-frequency cyclic force (Fig. 1.3) was associated with segmented chip formation and it was increased with increasing depth of cut and feed rate, and decreased with increasing cutting speed. To ascertain the effect of the tool entering angle on tool vibration and thus on tool life during milling titanium alloy, a cutting -force-based vibration analysis was presented in [75]. It was found that the directional cutting forces are widely affected by the tool entering angles used. The effects of cutting edge radius on cutting forces in orthogonal turning of Ti–6Al–4V titanium alloy under different levels of cutting speed and feed rate is reported in [74]. It was pointed out that the effect of cutting speed on the cutting force is non-linear and it depends on the cutting edge radius used.

For cutting forces reduction, down milling operations is mostly preferred than up milling during milling titanium alloys [51]. Furthermore, to reduce the machining costs, many alternative methods such as laser-assisted machining (LAM) [13, 16, 76–78], ultrasonic machining (USM) and electrical discharge

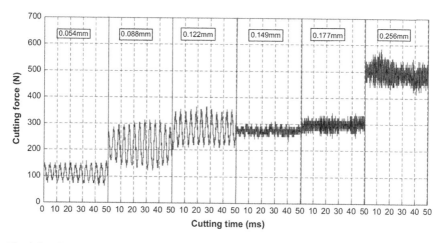

Fig. 1.1 Effect of feed rate on cutting force at cutting speed 75 m $\times$ min^{-1}

machining (EDM) [79–83] and consequently hybrid machining processes are proposed. As reported in [73], LAM and hybrid machining can reduce the cutting forces and provide substantial improvements in machinability of titanium alloys.

1.2.4 Chip Formation

A significant amount of experimental and modeling works reported the mechanism and morphology of chip formation during machining of titanium alloys [12, 19, 51, 69, 84–92]. The effects of broad range of cutting parameters, including feed rate, cutting speed and depth of cut and machining strategy and lubrication conditions on chip formation were studied. Amongst other machining factors, high pressure coolants (HPCs) have also important effects on improving the chip formation and consequently tool life, when machining titanium alloys [19, 51, 55, 93, 94]. Both continuous and segmented chip formation processes were observed in [28] under low cutting speed and large feed rate. According to [51], the chips in titanium alloys are generally thin and segmented. It is believed that the main reasons of segmented chip formation are due to either (1) the growth of cracks from the outer surface of the chip [95, 96] or (2) adiabatic shear band formation which is caused by the localized shear deformation resulting from the predominance of thermal softening over strain hardening [97, 98]. Gente and Hoffmeiste [84] reported the chip formation of Ti–6Al-4V at very high cutting speed, ranging between 30 m $\times$ min^{-1} and 6,000 m $\times$ min^{-1} (see Figs. 1.2, 1.3). According to experimental results, the structure of segmentation was changed at the cutting speed exceeding 2,000 m $\times$ min^{-1}. Furthermore; no change in specific cutting energy coincides with this change in structure [84]. Palanisamy et al. [19] studied the

Fig. 1.2 Schematic diagram
of segmented chip formation
when machining titanium
alloys [84]

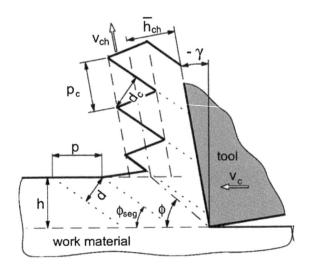

effects of HPC in machining of titanium alloys and observed that longer tool life
and better surface finish were resulted when using HPC (Figs. 1.4, 1.5). The
microstructures and morphology of chip formation of 4340 steel, 6061-T6 aluminium alloy and Ti–6Al-4V titanium alloy were examined during orthogonal
cutting tests under various levels of cutting speed [86]. The chip formation in
Ti–6Al-4V titanium alloys appeared to be segmented at all speed, but it became
continuous macroscopically at high speeds.

The chip formation in machining titanium alloys is strongly influenced by the
microstructural state of the material [90, 92]. The feasibility of dry cutting process
in machining Ti–6Al-4V titanium alloys with cemented tungsten carbide tools was
investigated in [91]. The rake angles 0, 15 and 30° as well as three cutting speeds
(15, 30 and 60 m × min^{-1}) and three feeds (0.1, 0.2 and 0.3 mm x rev^{-1}) were
used and morphology of chip formation, cutting force components and tool wear
were recorded. A significant correlation was found between the evolution of
cutting forces, the tool damage modes and the work part surface roughness, which
all indicate the significant influence of the cutting parameters and the insert
geometry on the tool wear and on the quality of the finished surface (Fig. 1.6).

Several analytical and numerical modeling of chip formation mechanism and
morphology during machining titanium alloys are reported in literature [69, 87–
89], but none can adequately predict the chip formation profile. Experimental and
analytical investigations of the shear localization phenomenon in orthogonal
machining of commercially pure titanium alloys were reported in [89], and a
thermo-plastic criterion for the onset of localization was presented. It was found
that in a wide range of cutting speeds used, the flow localization in the machining
of CP titanium is thermo-plastic. It was observed that at low cutting speeds, micro
cracks preceded flow instability. Hua and Shivpuri [88] presented a new

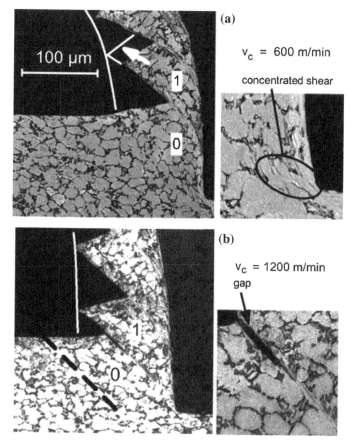

Fig. 1.3 Quick stop photomicrographs at different stages of titanium chip formation [84]

Fig. 1.4 Chips obtained after 1 min cutting time (first cut) from the application of **a** standard pressure coolant and **b** HPC [19]

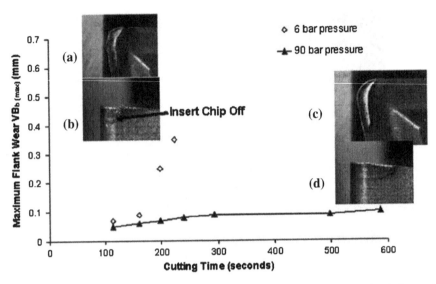

Fig. 1.5 Tool wear along with **a** crater view and **b** flank view of the insert obtained after tool chip-off with the application of standard pressure coolant. **c** crater view and **d** flank view of the insert [19]

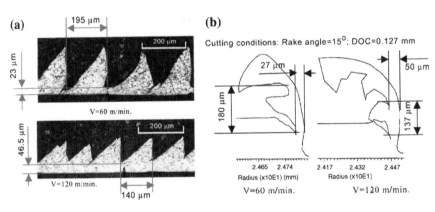

Fig. 1.6 Titanium chip morphology: **a** experimental results; **b** simulation results [88]

interpretation of chip segmentation in orthogonal machining of Ti–6Al-4V, based on an implicit, Lagrangian non-isothermal rigid–viscoplastic FE simulation, where a dynamic flow stress model based on high strain rate and high temperature, and a ductile fracture criterion based on the strain energy were applied to crack initiation during the chip segmentation process. The modeling results correlated well with experimental values (Fig. 1.7). Generally segmented chips occur during machining titanium alloys under wide range of cutting speeds and feed rates. This phenomenon is considered as a major concern during machining titanium alloys. Thus, it is

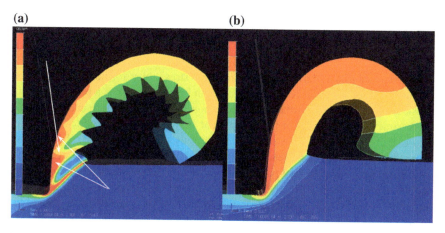

Fig. 1.7 Temperature distribution at the tool–chip interface for **a** a segmented chip; **b** a continuous chip [69]

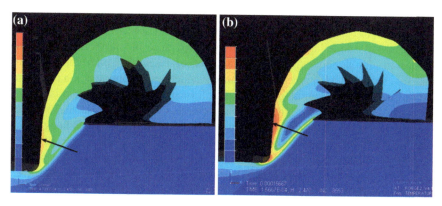

Fig. 1.8 Temperature distribution at the tool–chip interface under a cutting speed: **a** 60 m × min^{-1} and **b** 180 m × min^{-1} [69]

very much encouraged to predict the chip formation morphology before running the cutting operation. Calamaz et al. [69] developed a new material constitutive law model to predict the chip formation and shear localization when machining titanium alloys. The effects of strain, strain rate, temperature on the chip formation were studied in this work (Figs. 1.7, 1.8). The effects of two different strain softening levels and machining parameters on the cutting forces and chip morphology were studied. The predicted chip morphology, cutting and feed forces correlated well with experimental results.

1.2.5 Other Main Machinability Criteria: Residual Stress, Burr Formation and Particle Emission

Other machinability criterions include residual stress, burr formation and particle emission. The presence of residual stresses can be considered as a potential source of risk in terms of crack initiation, propagation and fatigue failure and effectively has detrimental influences on the component life. It has been reported in the literature that the residual stresses that commonly occur are tensile [67, 99] or compressive [100]. Different results can be found in literature with respect to effects of cutting parameters and tool parameters on residual stress recorded in machining titanium alloys. Furthermore, the presence and grade of compressive peak residual stresses [72, 99, 100], as well as the depth where residual stresses are levelled [72] are also not agreed upon [58]. The full description of machining induced surface integrity in titanium alloys is not within the scope of the current work. For more details in this regard, interested readers are encouraged to refer to [58].

Through growing demands on part quality, functional performance and global competition, special attention has been paid to burr formation which appears to be one of the major troublesome impediments to part quality, high productivity and automation [101]. Significant amount of research works have been conducted on understanding, modeling, minimization and avoidance of burr formation during various machining operations. Great amount of research works conducted by Niknam et al. [102-112] on characterization, modeling and optimization of milling and drilling burrs in aluminium alloys. Burr formation in other commercially used aerospace and automotive alloys such as, stainless steel, steel and nickel based alloys [113-120] have been extensively reported in the literature. Although relatively less amount of work reported the burr formation in titanium alloys [99, 121, 122], but it could be inferred that the irrespective to machining and work part used, keys points as already pointed out by Aurich et al. [120] remain to date; they included: (I) development of databases for adequate selection of optimal cutting conditions; (II) developing link between the burr size and deburring difficulty, and (III) improving and automating burr detection and characterization strategies.

A case study on metallic particle emission when machining titanium is presented in Chap. 2.

1.3 Microstructural Aspects During Machining of Titanium Alloys

Titanium alloys are widely used in different industries that require excellent mechanical resistance at high temperatures and resistance to corrosion. However, machining of titanium alloys presents some difficulties caused by the particularities in term of metallurgical properties [123, 124].

Fig. 1.9 Microstructure of
the Ti-6Al-4V alloy [129]

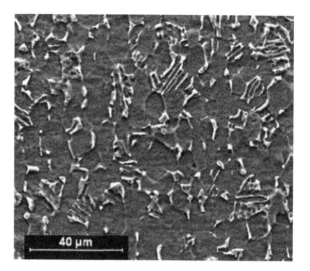

As discussed in [25, 32], the thickness of the very fine plastically deformed
layer formed in the sub-surface of the titanium alloys during dry turning operation
is affected by the cutting speed and feed rate [25, 32]. This layer is located in the
sub-machined surface, so called white layer, has been also observed by other
researchers [125–127]. However, at very high-speed dry turning of Ti-6Al-4V, the
white layer was no observed [126, 127]. These works present a change in the grain
direction in line with the tool pass, but no phase transformation in the machined
surface exist, even when using a wide range of cutting speed. However, a particular thin zone (thickness from 0.5 to 5 μm) was observed near to the machined
surface with grains quasi-parallel to the cutting direction [126–129].

As discussed earlier, the Ti-6Al-4V is the most highly used titanium alloy, which
can be generally found in two phases α and β. The microstructure of the Ti-6Al-4V
alloy is depicted in Fig. 1.9 [129]. The high temperatures generated during cutting
caused by the low thermal conductivity of the Ti-6Al-4V alloys can produce a
softened zone under the surface [129]. It has been found that the roughing affects
the microstructure of the machined surface but not significantly compared to the
finishing pass that defines the thermo mechanical state of the real surface [129].

1.4 Effects of Lubrication/Coolant Modes

The use of metal working fluids (MWFs) is widely reported in industry for decades,
despite the facts that lubricated machining is associated with environmental concerns such as airborne mist, smoke and other particulates in shop air quality as well
as massive operating and disposal costs. The increasing demands for environmentally friendly machining operations urged researchers to propose new

alternative methods to replace cutting fluids. Thus, dry cutting, high pressure cooling (HPC), minimum quantity lubrication (MQL) and cryogenic cooling became the primary source of interests in this regard [17, 51, 52, 55, 58, 60, 73, 93, 94, 130–135], which all aim to improve the machinability of the titanium alloys. As pointed out in [51], the thin and segmented chips in titanium alloys have high edge temperatures, that in fact making the correct selection of cooling method. Rotella et al. [60] presented a comparative study of machining Ti–Al-4V titanium alloy under dry, MQL and cryogenic conditions. It was found that the cooling conditions could affect the surface integrity. In particular, cryogenic method significantly improved the product surface. The grindability of Ti-6A1-4V titanium alloy with MQL when using synthetics oils and vegetable oils was evaluated in [130]. It was respectively found that MQL grinding led to better results compared to that observed in conventional grinding methods. MQL grinding with synthetics oil presented better surface quality, better surface morphology and lower grinding forces than vegetable oils, whereas vegetable oils has shown better cooling effects in grinding zone [130]. Furthermore, vegetable oil reduces the health hazard factors [136]. As reported in [134], the MQL technique reduces the power and specific energy compared to conventional soluble oil. Experimental studies were conducted on high speed end milling of Ti-6-A1-4V titanium alloy under various cooling/lubrication conditions to find the optimal condition to improve the tool life [135].

The titanium alloy is generally cut at low cutting speed with the cutting fluid in order to prevent severe tool wear [52]. According to experimental works reported in [52], the titanium alloy was cut (milled) in the dry and mist methods to avoid the thermal effects on the inserts. The wear progress in mist lubrication conditions was longer than that in dry. The longer tool life was recorded in mist cutting condition compared to that observed in dry cutting condition. As reported by Rahman et al. [132], for continuous cutting operations, dry cutting has shown successful performance at low cutting speed and MQL is more cost effective approach compared to flood condition. The effects of droplet spray behaviour on machining performance, tool wear/life and surface roughness during turning of Ti-6A1-4V titanium alloy was reported in [133]. It was found that tool life can be significantly improved at higher droplet velocity, a lower gas velocity and a longer spray distance. In fact, the high HPC not only enhances the production efficiency, but also improves the chip removal, resulting to better tool life during machining titanium alloys [19, 55, 93, 94]. As shown in [19], the tool life was significantly improved when using coolant at high pressure turning of Ti–Al-4V titanium alloy. The HPC was applied on the rake face and on the flank face of the cutting tool in face-grooving operations of Ti-6Al-4V titanium alloy, which are considered as common and time-consuming turning operations in the aerospace industry [93]. Due to the geometry of the groove and those chips that obstruct the coolant flow, cooling effect becomes low in grooving operations. It was found that when high-pressure rake face cooling is applied, the workpiece surface may be adversely affected by the chip flow. Applying HPC to the tool flank face increases tool life by

50–100 % relative to conventional flood flushing [93]. In addition to reported cooling methods, particular attention has been paid to cryogenic cooling methods. In particular, liquid nitrogen (LN2) has been widely used as cryogenic coolant for machining titanium alloys [17, 137, 138].

The strength and hardness of machined titanium specimens are enhanced under cryogenic conditions while toughness and ductility tend to reduce at decreased temperature [138]. According to [17], as the nitrogen evaporates, a nitrogen cushion formed by evaporating nitrogen lowers the coefficient of friction between the chip and the tool. It is to underline that since lubricated machining is often required to produce high quality products, the complete removal of cutting fluids is not always possible, especially when hard-to-cut materials, such as titanium alloys are machined [131]. However, the correct selection of the lubrication condition depends on many factors, including the material being machined, the type of cutting tool employed, the machining conditions as well as the surface finishing, dimensional quality, and shape of the product.

1.5 Case Study: Metallic Particle Emission During Turning of Ti-6A-14V Titanium Alloy

.1.6 Introduction

Micro particles (generally PM2.5) and ultrafine particles, also called nanoparticles are considered as a serious health problem. The fine particles are inhalable and can affect the respiratory system. The behavior of nanoparticles is complicated and their toxicity is not still well understood. Consequently, nanoparticles are considered as potentially dangerous. The means of contamination are various. The nanoparticles can generally penetrate into body through three main ways: inhalation, skin and digestive [2]. Research works carried out on the toxicity of the nanoparticles have shown that the TiO_2 causes the acute skin irritation [2]. Because of the danger posed by fine and ultrafine particles produced during machining processes, particle emissions now constitute a performance criterion ranking at the same level as other traditional criteria such as productivity, cycle time, precision, part quality and manufacturing costs [139].

Machining processes can produce liquid or solid aerosols of different sizes. The use of cutting fluids generates liquid aerosols which can be harmful to the operator and the environment [62, 140–148]. Wet machining has been found to generate more fine and ultrafine particles than dry machining [149, 150]. Although, dry machining eliminates the cutting fluid, but it might reduce the tool life and deteriorate the surface finish and increases the power requirements. However, dry machining can be advantageous if the optimal cutting parameters and tooling are selected. Experimental and simulation works have shown that dry aerosols generated during metal machining depend on the work material, cutting condition, as well as the cutting parameters [21, 150, 151] and the tool geometry [152–156]. The chip formation mode and the particle emission mechanisms of titanium alloys are different as compared to other materials. Furthermore, the titanium and aluminum alloys are both used in aeronautic and automotive industries. The comparison between the machining of titanium and aluminum alloys is very challenging due to the high speed threshold value that is different for both materials. In this study, particle emission and chip formation mode represent the main elements that were investigated. This study presents the new sources and mechanisms of particle emissions during machining processes.

1.7 Experimental Plan

The cutting conditions used for the experiments are summarized in Table 1.2. The particle concentration was measured using a laser photometer (TSI8520 Dustrack) equipped with a filter allowing the selection of the particle size. In this study, the particles investigated had an aerodynamic diameter smaller than 2.5 mm (PM2.5). The Dustrack was connected to the dust recovery enclosure by a suction pipe, and the air flow rate used during sampling was around $1.7l \times min^{-1}$ (Fig. 1.10). Several parameters can be used in the machining processes, and the removed chip quantity varies according to the process and the shape of the part being machined. Therefore, it is important to use an appropriate parameter to evaluate the particle emission. This parameter can be related to the cutting parameter, the working process and/or the volume of chips removed. Khettabi et al. [152] developed a new dimensionless index, so called "the Dust Unit (Du)", defined as the ratio of the dust mass (m_{Dust}) to the quantity of chip removed (m_{Chip}) from the workpiece material (Eq. 1.1). The dust mass is calculated according to the dust concentration, the particle sampling flow rate, the particle density and the sampling time [152]. The particle emission was estimated using the mass concentration recorded by Dustrack and subsequently transformed to the Dust Unit (Du).

$$\text{Dust unit} = \text{Du} = \frac{m_{\text{dust}}}{m_{\text{chip}}} \tag{1.1}$$

1.8 Results

Figures 1.11 and 1.12 display the effects of tool cutting edge angle (κ) and cutting speeds on metallic particle emission. According to [152], the tool cutting edge angle (κ) can affect the particle emission in machining of steel and aluminum alloys. For the titanium alloy Ti-6A1-4V, the behavior observed with respect to the influence of the tool cutting edge angle on particle emission appears to be similar

Table 1.2 Experimental parameters used

Operation	Turning
Feed rate (mm × rev^{-1})	0.1
Depth of cut (mm)	0.5
Cutting speed (m × min^{-1})	50–300
Tool material	Uncoated carbide
Tool cutting edge angle κ(°)	70–110
Lubricant and coolant	None
Work materials	Titanium alloy Ti-6Al-4V

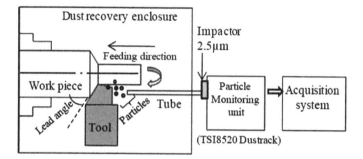

Fig. 1.10 Experimental setup used for the particle emission study

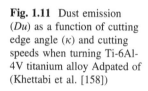

Fig. 1.11 Dust emission (Du) as a function of cutting edge angle (κ) and cutting speeds when turning Ti-6Al-4V titanium alloy Adpated of (Khettabi et al. [158])

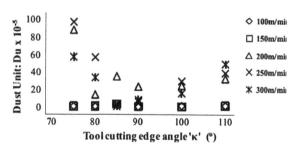

to that of steels. The minimum particle emission was found with a tool cutting edge angle (κ) of 90°. However, with tool cutting edge angle (κ) greater or less than 90°, particles increases (Fig. 1.11), thus confirming the fact that a range of tool cutting edge angles (κ) exist over which dust emission becomes minimal. Based on reported works in [152–154], the edge angles (κ) should be within the range of 90 ± 10°. While comparing this result with the one obtained on aluminium alloys, Khettabi et al. [2] found that the machining of the Ti-6Al-4V titanium alloy generates more particles than compared to the aluminum alloy by a factor of about 20. There are two in peaks of particle emission when turning the Ti-6Al-4V titanium alloy (Fig. 1.12), one being at low cutting speeds range and

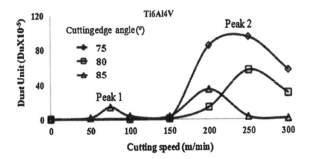

Fig. 1.12 Influence of the cutting speed and tool cutting edge angle (κ) on dust emission when turning Ti-6Al-4V titanium alloy [158]

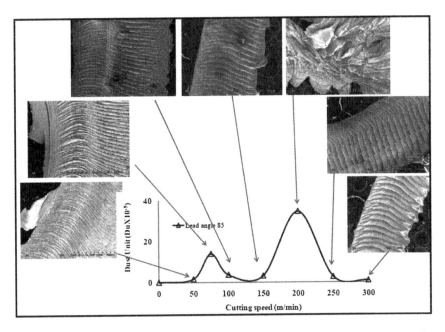

Fig. 1.13 Behavior of dust emission and chip formation mode during turning of titanium alloy Ti-6Al-4V when using different cutting speeds [158]

another at higher speeds. These two peaks correspond to different chip formation modes as noticeable in Fig. 1.13. At $200 \text{ m} \times \text{min}^{-1}$, the chip form is completely different compared to the other cutting speeds used. This could be due to special mechanical behaviour of Ti-6Al-4V titanium alloy at high cutting speed (Fig. 1.13). Consequently, the particle emission should be different at this range of cutting speed. This could be due to special the maximum particle emission as a function of cutting speed presents only one range of speeds [152–156]. Also, regarding the aluminum alloys, it has been found that the particle emission

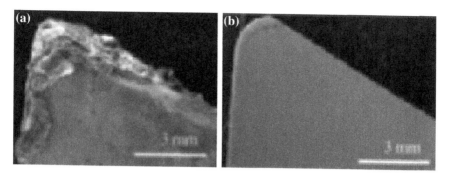

Fig. 1.14 Wear observed on tool after machining titanium alloy (Ti-6A1-4V) and aluminum alloy (6061-T6) at a cutting speed of 200 m × min^{-1} for the same time period Adpated of [158]

decreases relatively when the cutting speed exceeds 300 m × min^{-1} [153]. However, the speed range of 80–100 m × min^{-1} is considered as high speed machining when dealing with the titanium alloys [36, 157]. Therefore, cutting tests on Ti-6A1-4V titanium alloy were performed at speeds below 100 m × min^{-1}.

Since the similar cutting tool was used in the tests, it was found that the tool wears out quicker during the machining of the titanium alloy (Ti-6A1-4V) compared to that observed in aluminum alloy (Fig. 1.13). Figure 1.13 compares the state of the tools in order to show the difficulty of machining titanium alloy. The tools shown in Fig. 1.13a and c were used for the same time period. During the machining of the titanium alloy, it was observed that the tool cannot resist at the cutting speed of 200 m × min^{-1}, and instead deteriorated. At this critical value of cutting speed, the chip becomes very soft and sticks to the tool, reflecting the excessive temperature rise (Fig. 1.14a, b). According to the materials tested, this phenomenon was only observed during the machining of the titanium alloy (Ti-6A1-4V), while the cutting tool remained noticeably unworn during machining of 6061-T6 aluminum alloy at the cutting speeds tested (Fig. 1.14a, b).

1.9 Discussion

Dislike aluminium alloys, machining of titanium alloys are associated with high tool vibration, different chip formation mode (Figs. 1.11, 1.12, 1.13, 1.14) and short tool life [1, 3–5]. This is due to the special mechanical properties and the behavior of the alloys during machining processes. At low cutting speed, below 150 m × min^{-1}, the variation of particle emission of Ti-6Al-4V titanium alloy seems similar to that of the 6061-T6 aluminum alloy. At this cutting condition, the mechanism of particle emission is affected by energy approach, combined with macroscopic friction (tool-chip), microfriction, and plastic deformation of materials.

The energy provided by the cutting tool must be higher than the extraction energy required for a particle to leave the parent material. The wear observed on the tools used during the machining of the titanium alloy presents a high level of degradation that itself is an evidence of a high level of particle emissions. Above 150 m $\times$ min^{-1}, the titanium alloy presents different modes of particle emission (Figs. 1.13 and 1.14). At the cutting speed 200 m $\times$ min^{-1}, which is considered as a critical cutting speed, the chips become very soft and stick to the tool. This sticking phenomenon during the machining of titanium alloys occurs only at very high temperatures. However, during the machining of the 6061-T6 aluminum alloy, the tool remained almost sharp at all tested speeds (Fig. 1.14b). In general, high speed machining of the titanium alloy occurs at 80–100 m $\times$ min^{-1}. To use a higher level of speed (e.g. 150 m $\times$ min^{-1}), very high performance tool materials such as CBN might be required [36, 157]. The cutting speed of 300 m $\times$ min^{-1} or higher during the machining of titanium alloys is only allowed under special finishing conditions using CBN tools [36, 157].

The presence of two particle emission peaks as a function of cutting speed is typical for the machining of titanium alloys (Figs. 1.12, 1.13). At speeds lower than 100 m $\times$ min^{-1}, the particle emission behavior of titanium alloys is similar to that of aluminum alloys. However, above 150 m $\times$ min^{-1}, the mechanical behavior of titanium is changed and the emission sources enlarged (Table 1.3 and Fig. 1.13). The chip mode presents a good index for the mechanisms of particle emissions during machining operations. According to Fig. 1.12 and the Table 1.3, the chip form is similar at the two ranges of cutting speed (100–150 m $\times$ min^{-1} and 250–300 m $\times$ min^{-1}). At these two ranges, the chip segmentation density seems similar and the particle emissions were very low compared to the two peaks at the cutting speeds of 75 m $\times$ min^{-1} and 200 m $\times$ min^{-1}. At the first peak, the chip segmentation density is very high which increases the particle emissions. At very high cutting speeds (second peak), the titanium alloy becomes reactive and large particles quickly decay by oxidation and explosion processes. The chip mode is also modified at very high cutting speeds, and the tool wear rate is accelerated (Table 1.3 and Fig. 1.13), and therefore, another phenomenon is triggered after a certain cutting speed during the machining of titanium alloys.

In conclusion, the study of particle emissions during the machining of Ti-6Al-4V titanium alloy shows that the chip formation mode is a good indicator of particle emission for the tested titanium alloy. The behavior of the Ti-6Al-4V titanium alloy during machining operations strongly affects the mechanisms of particle emission. The disintegration of coarse particles produced during the machining of titanium alloys under certain conditions produces a number of fine and ultrafine particles. Machining titanium alloys at very high speed can be beneficial from a particle emissions point of view, but it requires special tools and cutting conditions.

Table 1.3 Chip formation during dry machining of Ti-6A1-4V titanium alloy [158]

Cutting speed (m × min⁻¹)	Low magnification 100×	High magnification 1500×
75		
100		
200		
250		

1.10 Conclusion

Despite recent developments and extensive usage of titanium alloys, machining of titanium alloys still remains as a major industrial concern: short tool life, low metal removal rates, higher cutting force and temperature, and poor surface quality.

To improve the machinability of titanium alloys, special attentions must be paid to machining strategies and cutting tools. There are many types of cutting tools employed for machining of titanium alloys. Amongst, carbide tools are still the most commonly used materials. The use of coated tools does not show a considerable improvement on the machinability of titanium alloys. The cutting temperature and high pressure at the tool-chip interface, built up edge (BUE) formation and the chemical interaction between the titanium and the tool are the main reasons of the tool wear. In fact, the tendency of titanium alloys to react with the most of cutting tool materials is the main factor of tool wear which hinders the machinability of titanium alloys. Tool wear mostly occur in the tool flank side in both coated and uncoated tools in machining titanium alloys.

The chip formation in titanium alloys were found as (1) segmented, associated with a cyclic cutting force, and (2) continuous under static force at low cutting speed and large feed rate under turning operations. According to literature, new physical model need to be developed to better explain the chip formation morphology in titanium alloys.

Low surface quality is resulted when machining titanium alloys; the high temperature during machining of titanium alloys is the main reason for high surface roughness values. Also, the built up edge (BUE) takes place and increases the roughness values.

Machining titanium alloys is generally associated with high cutting forces. Increased cutting forces are also attributed to strain rate hardening. Substantial improvements in decreasing the cutting forces are observable when using hybrid machining approaches, such as laser-assisted machining (LAM). The behavior of the Ti-6A1-4V titanium alloy during machining operations strongly affects the mechanisms of particle emission and the chip formation. The disintegration of coarse particles produced during the machining of titanium alloys under certain conditions produces fine and ultrafine particles. Machining Ti-6A1-4V titanium alloy at high speed can be beneficial from a particle emission point of view, but it requires special tools.

According to [51], the general machining recommendation for titanium alloys are: (1) using sharp cutting edge tool (2) providing well-clamped work parts for stable cutting conditions(3) applying appropriate cooling methods (Wet, MQL, Cryogenic, etc.), and (4) minimising the vibration tendencies.

References

1. Ezugwu E, Wang Z (1997) Titanium alloys and their machinability—a review. J Mater Process Technol 68:262–274. doi:10.1016/S0924-0136(96)00030-1
2. Khettabi R, Fatmi L, Masounave J, Songmene V (2013) On the micro and nanoparticle emission during machining of titanium and aluminum alloys. CIRP J Manuf Sci Technol 6:175–180. doi:10.1016/j.cirpj.2013.04.001
3. Ezugwu E, Bonney J, Yamane Y (2003) An overview of the machinability of aeroengine alloys. J Mater Process Technol 134:233–253. doi:10.1016/S0924-0136(02)01042-7
4. Hughes J, Sharman A, Ridgway K (2006) The effect of cutting tool material and edge geometry on tool life and workpiece surface integrity. Proc Inst Mech Eng Part B J Eng Manuf 220:93–107. doi:10.1243/095440506X78192

5. Sun J, Guo Y (2009) A comprehensive experimental study on surface integrity by end milling Ti–6Al-4V. J Mater Process Technol 209:4036–4042. doi:10.1016/j.jmatprotec. 2008.09.022
6. Hong H, Riga AT, Gahoon JM, Scott CG (1993) Machinability of steels and titanium alloys under lubrication. J Wear, Part A 162–164:34–39. doi:10.1016/0043-1648(93)90481-Z
7. Ohkubo C, Watanabe I, Ford JP, Nakajima H, Hosoi T, Okabe T (2000) The machinability of cast titanium and Ti–6Al-4V. J Biomaterials 21:421–428. doi:10.1016/S0142-9612(99)00206-9
8. Watanabe I, Kiyosue S, Ohkubo C, Aoki T, Okabe T (2002) Machinability of cast commercial titanium alloys. J Biomed Mat Res 63:760–764. doi:10.1002/jbm.10413
9. Rahman M, Wong YS, Zareena AR (2003) Machinability of titanium alloys. JSME Int J Ser C 46:107–115. doi:10.1299/jsmec.46.107
10. Arrazola PJ, Garay A, Iriarte LM, Armendia M, Marya S, Le Maître F (2009) Machinability of titanium alloys (Ti6Al4V and Ti555.3). J Mater Process Technol 209:2223–2230. doi:10.1016/j.jmatprotec.2008.06.020
11. Kikuchi M (2009) The use of cutting temperature to evaluate the machinability of titanium alloys. Acta Biomater 5:770–775. doi:10.1016/j.actbio.2008.08.016
12. Rashid RR, Sun S, Wang G, Dargusch M (2011) Machinability of a near beta titanium alloy. Proc Inst Mech Eng Part B J Eng Manuf 225:2151–2162. doi:10.1177/2041297511406649
13. Dandekar CR, Shin YC, Barnes J (2010) Machinability improvement of titanium alloy (Ti–6Al-4V) via LAM and hybrid machining. Int J Mach Tools Manuf 50:174–182. doi:10.1016/j.ijmachtools.2009.10.013
14. Khanna N, Garay A, Iriarte LM, Soler D, Sangwan KS, Arrazola PJ (2012) Effect of heat treatment conditions on the machinability of Ti64 and Ti54M alloys. Procedia CIRP 1:477–482. doi:10.1016/j.procir.2012.04.085
15. Khanna N, Sangwan KS (2013) Comparative machinability study on Ti54M titanium alloy in different heat treatment conditions. Proc Inst Mech Eng Part B J Eng Manuf 227:96–101. doi:10.1177/0954405412466234
16. Yang JHN, Brandt M, Sun SJ (2009) Numerical and experimental investigation of the heat-affected zone in a laser-assisted machining of Ti-6Al-4V alloy process. Mater Sci Forum 618–619:143–146. doi:10.4028/www.scientific.net/MSF.618-619.143
17. Hong SY, Ding Y (2001) Cooling approaches and cutting temperatures in cryogenic machining of Ti-6Al-4V. Int J Mach Tools Manuf 41:1417–1437. doi:10.1016/S0890-6955(01)00026-8
18. Hong SY, Markus I, Jeong W (2001) New cooling approach and tool life improvement in cryogenic machining of titanium alloy Ti-6Al-4V. Int J MachTools Manuf 41:2245–2260. doi:10.1016/S0890-6955(01)00041-4
19. Palanisamy S, McDonald SD, Dargusch MS (2009) Effects of coolant pressure on chip formation while turning Ti6Al4V alloy. Int J Mach Tools Manuf 49:739–743. doi:10.1016/j.ijmachtools.2009.02.010
20. An QL, Fu YC, Xu JH (2011) Experimental study on turning of TC9 titanium alloy with cold water mist jet cooling. Int J Mach Tools Manuf 51:549–555. doi:10.1016/j.ijmachtools.2011.03.005
21. Balout B, Songmene V, Masounave J (2007) An experimental study of dust generation during dry drilling of pre-cooled and pre-heated workpiece materials. J Manuf Processes 9:23–34. doi:10.1016/S1526-6125(07)70105-6
22. Khanna N, Sangwan K (2010) Cutting Tool Performance in Machining of Ti555. 3, Timetal® 54M, Ti 6-2-4-6 and Ti 6-4 Alloys: A Review and Analysis. In: Proceedings of 2nd CIRP international conference on process machine interactions, June 10–11, 2010, Vancouver
23. Rahim EA, Sharif S (2006) Investigation on tool life and surface integrity when drilling Ti-6Al-4V and Ti-5Al-4V-Mo/Fe. JSME Int J Ser C 49:340–345. doi:10.1299/jsmec.49.340

24. Chen L, El-Wardany T, Harris W (2004) Modelling the effects of flank wear land and chip formation on residual stresses. CIRP J Manuf Sci Technol 53:95–98. doi:10.1016/S0007-8506(07)60653-2

25. Che-Haron C, Jawaid A (2005) The effect of machining on surface integrity of titanium alloy Ti–6 %Al–4 %V. J Mater Process Technol 166:188–192. doi:10.1016/j.jmatprotec.2004.08.012

26. Wang Z, Rahman M, Wong Y (2005) Tool wear characteristics of binderless CBN tools used in high-speed milling of titanium alloys. J Wear 258:752–758. doi:10.1016/j.wear.2004.09.066

27. Amin A, Ismail AF, Nor Khairusshima M (2007) Effectiveness of uncoated WC–Co and PCD inserts in end milling of titanium alloy—Ti–6Al–4V. J Mater Process Technol 192:147–158. doi:10.1016/j.jmatprotec.2007.04.095

28. Sun S, Brandt M, Dargusch M (2009) Characteristics of cutting forces and chip formation in machining of titanium alloys. Int J Mach Tools Manuf 49:561–568. doi:10.1016/j.ijmachtools.2009.02.008

29. Thomas M, Turner S, Jackson M (2010) Microstructural damage during high-speed milling of titanium alloys. Scr Mater 62:250–253. doi:10.1016/j.scriptamat.2009.11.009

30. Ginting A, Nouari M (2007) Optimal cutting conditions when dry end milling the aeroengine material Ti–6242S. J Mater Process Technol 184:319–324. doi:10.1016/j.jmatprotec.2006.10.051

31. Ginting A, Nouari M (2009) Surface integrity of dry machined titanium alloys. Int J Mach Tools Manuf 49:325–332. doi:10.1016/j.ijmachtools.2008.10.011

32. Che-Haron CH (2001) Tool life and surface integrity in turning titanium alloy. J Mater Process Technol 118:231–237. doi:10.1016/S0924-0136(01)00926-8

33. Sridhar B, Devananda G, Ramachandra K, Bhat R (2003) Effect of machining parameters and heat treatment on the residual stress distribution in titanium alloy IMI-834. J Mater Process Technol 139:628–634. doi:10.1016/S0924-0136(03)00612-5

34. Mantle A, Aspinwall D (1997) Surface integrity and fatigue life of turned gamma titanium aluminide. J Mater Process Technol 72:413–420. doi:10.1016/S0924-0136(97)00204-5

35. Mantle A, Aspinwall D (2001) Surface integrity of a high speed milled gamma titanium aluminide. J Mater Process Technol 118:143–150. doi:10.1016/S0924-0136(01)00914-1

36. Zoya Z, Krishnamurthy R (2000) The performance of CBN tools in the machining of titanium alloys. J Mater Process Technol 100:80–86. doi:10.1016/S0924-0136(99)00464-1

37. Kitagawa T, Kubo A, Maekawa K (1997) Temperature and wear of cutting tools in high-speed machining of Inconel 718 and Ti6Al 6V2Sn. J Wear 202:142–148. doi:10.1016/S0043-1648(96)07255-9

38. Nabhani F (2001) Machining of aerospace titanium alloys. Rob Comput Integr Manuf 17:99–106. doi:10.1016/S0736-5845(00)00042-9

39. Cui W, Jin Z, Guo A, Zhou L (2009) High temperature deformation behavior of α+β-type biomedical titanium alloy Ti–6Al–7Nb. Mater Sci Eng A 499:252–256. doi:10.1016/j.msea.2007.11.109

40. Karpat Y (2011) Temperature dependent flow softening of titanium alloy Ti6Al4V: an investigation using finite element simulation of machining. J Mater Process Technol 211:737–749. doi:10.1016/j.jmatprotec.2010.12.008

41. Sun S, Brandt M, Dargusch M (2010) The effect of a laser beam on chip formation during machining of Ti6Al4V alloy. Metall Mater Trans A 41:1573–1581. doi:10.1007/s11661-010-0187-5

42. Kahles J, Field M, Eylon D, Froes F (1985) Machining of titanium alloys. JOM 37:27–35

43. Honnorat Y (1996) Issues and breakthrough in the manufacture of turboengine titanium parts. Mater Sci Eng A 213:115–123. doi:10.1016/0921-5093(96)10229-X

44. Jawaid A, Sharif S, Koksal S (2000) Evaluation of wear mechanisms of coated carbide tools when face milling titanium alloy. J Mater Process Technol 99:266–274. doi:10.1016/S0924-0136(99)00438-0

45. Ghani JA, Che Haron CH, Hamdan SH, Md Said AY, Tomadi SH (2013) Failure mode analysis of carbide cutting tools used for machining titanium alloy. Ceram Int 39:4449–4456. doi:10.1016/j.ceramint.2012.11.038

46. Haron C, Ginting A, Arshad H (2007) Performance of alloyed uncoated and CVD-coated carbide tools in dry milling of titanium alloy Ti-6242S. J Mater Process Technol 185:77–82. doi:10.1016/j.jmatprotec.2006.03.135

47. Prengel H, Pfouts W, Santhanam A (1998) State of the art in hard coatings for carbide cutting tools. Surf Coat Technol 102:183–190. doi:10.1016/S0257-8972(96)03061-7

48. Ezugwu E, Da Silva RR, Bonney J, Machado A (2005) The effect of argon-enriched environment in high-speed machining of titanium alloy.Tribol Trans 48:118–123. doi:10.1080/05698190590890290

49. M'saoubi R, Outeiro J, Changeux B, Lebrun J, Morao Dias A (1999) Residual stress analysis in orthogonal machining of standard and resulfurized AISI 316L steels. J Mater Process Technol 96:225–233. doi:10.1016/S0924-0136(99)00359-3

50. Liao Y, Shiue R (1996) Carbide tool wear mechanism in turning of Inconel 718 superalloy. J Wear 193:16–24. doi:10.1016/0043-1648(95)06644-6

51. Coromant S (1994) Modern metal cutting: a practical handbook: Sandvik Coromant, Sweden

52. Fujiwara J, Arimoto T, Tanaka K (2011) High speed milling of titanium alloy. Adv Mater Res 325:387–392. doi:10.4028/www.scientific.net/AMR.325.387

53. Zareena AR, Veldhuis SC (2012) Tool wear mechanisms and tool life enhancement in ultra-precision machining of titanium. J Mater Process Technol 212:560–570. doi:10.1016/j.jmatprotec.2011.10.014

54. Yang X, Richard Liu C (1999) Machining titanium and its alloys. J Mach Sci Tech 3:107–139. doi:10.1080/10940349908945686

55. Ezugwu E, Da Silva R, Bonney J, Machado A (2005) Evaluation of the performance of CBN tools when turning Ti–6Al-4V alloy with high pressure coolant supplies. Int J Mach Tools Manuf 45:1009–1014. doi:10.1016/j.ijmachtools.2004.11.027

56. Arunachalam R, Mannan M, Spowage A (2004) Residual stress and surface roughness when facing age hardened Inconel 718 with CBN and ceramic cutting tools. Int J Mach Tools Manuf 44:879–887. doi:10.1016/j.ijmachtools.2004.02.016

57. Arunachalam R, Mannan M, Spowage A (2004) Surface integrity when machining age hardened Inconel 718 with coated carbide cutting tools. Int J Mach Tools Manuf 44:1481–1491. doi:10.1016/j.ijmachtools.2004.05.005

58. Ulutan D, Ozel T (2011) Machining induced surface integrity in titanium and nickel alloys: a review. Int J Mach Tools Manuf 51:250–280. doi:10.1016/j.ijmachtools.2010.11.003

59. Crawforth P, Wynne B, Turner S, Jackson M (2012) Subsurface deformation during precision turning of a near-alpha titanium alloy. Scr Mater 67:842–845. doi:10.1016/j.scriptamat.2012.08.001

60. Rotella G, Dillon OW, Umbrello D, Settineri L, Jawahir IS (2014) The effects of cooling conditions on surface integrity in machining of Ti6Al4V alloy. Int J Adv Manuf Technol 71:47–55. doi:10.1007/s00170-013-5477-9

61. El-Wardany T, Kishawy H, Elbestawi M (2000) Surface integrity of die material in high speed hard machining, Part 1: micrographical analysis. J Manuf Sci Eng 122:620–631. doi:10.1115/1.1286367

62. Dhar N, Kamruzzaman M, Ahmed M (2006) Effect of minimum quantity lubrication (MQL) on tool wear and surface roughness in turning AISI-4340 steel. J Mater Process Technol 172:299–304. doi:10.1016/j.jmatprotec.2005.09.022

63. Ramesh S, Karunamoorthy L, Palanikumar K (2008) Surface roughness analysis in machining of titanium alloy. J Mater Manuf Process 23:174–181. doi:10.1080/10426910701774700

64. Kumar Pandey A, Kumar Dubey A (2012) Simultaneous optimization of multiple quality characteristics in laser cutting of titanium alloy sheet. Opt Laser Technol 44:1858–1865. doi:10.1016/j.optlastec.2012.01.019

65. Vijay S, Krishnaraj V (2013) Machining Parameters Optimization in End Milling of Ti-6Al-4V. Proc Eng 64:1079–1088. doi:10.1016/j.proeng.2013.09.186
66. Ribeiro MV, Moreira MRV, Ferreira JR (2003) Optimization of titanium alloy (6Al–4V) machining. J Mater Process Technol 143–144:458–463. doi:10.16/S0924-0136(03)00457-6
67. Özel T, Zeren E (2007) Finite element modeling the influence of edge roundness on the stress and temperature fields induced by high-speed machining. Int J Adv Manuf Technol 35:255–267. doi:10.1007/s00170-006-0720-2
68. Venugopal KA, Paul S, Chattopadhyay AB (2007) Tool wear in cryogenic turning of Ti-6Al-4V alloy. J Cryo 47:12–18. doi:10.1016/j.cryogenics.2006.08.011
69. Calamaz M, Coupard D, Girot F (2008) A new material model for 2D numerical simulation of serrated chip formation when machining titanium alloy Ti–6Al-4V. J Mach Tools Manuf 48:275–288. doi:10.1016/j.ijmachtools.2007.10.014
70. Thepsonthi T, Özel T (2013) Experimental and finite element simulation based investigations on micro-milling Ti-6Al-4V titanium alloy: Effects of CBN coating on tool wear. J Mater Process Technol 213:532–542. doi:10.1016/j.cirp.2011.03.087
71. Ulutan D, Özel T (2012) Methodology to determine friction in orthogonal cutting with application to machining titanium and nickel based alloys. Proceedings of ASME 2012 international manufacturing science and engineering conference. Notre Dame, June 4–8, 2012. doi:10.1115/MSEC2012-7275
72. Axinte D, Dewes R (2002) Surface integrity of hot work tool steel after high speed milling-experimental data and empirical models. J Mater Process Technol 127:325–335. doi:10.1016/S0924-0136(02)00282-0
73. Ezugwu E (2005) Key improvements in the machining of difficult-to-cut aerospace superalloys. Int J Mach Tools Manuf 45:1353–1367. doi:10.1016/j.ijmachtools.2005.02.003
74. Wyen CF, Wegener K (2010) Influence of cutting edge radius on cutting forces in machining titanium. CIRP Ann Manuf Technol 59:93–96. doi:10.1016/j.cirp.2010.03.056
75. Ítalo Sette Antonialli A, Eduardo Diniz A, Pederiva R (2010) Vibration analysis of cutting force in titanium alloy milling. Int J Mach Tools Manuf 50:65–74. doi:10.1016/j.ijmachtools.2009.09.006
76. Germain G, Morel F, Lebrun JL, Morel A (2007) Machinability and surface integrity for a bearing steel and a titanium alloy in laser assisted machining. Lasers Eng
77. Sun S, Brandt M (2007) Laser-assisted machining of titanium alloys. Industrial lasers report, IRIS Swinburne University of Technology, Melbourne
78. Zitoune R, Krishnaraj V, Davim JP (2013) Laser assisted machining of titanium alloys. Mater Sci Forum 763:91–106. doi:10.4028/www.scientific.net/MSF.763.91
79. Wansheng Z, Zhenlong W, Shichun D, Guanxin C, Hongyu W (2002) Ultrasonic and electric discharge machining to deep and small hole on titanium alloy. J Mater Process Technol 120:101–106. doi:10.1016/S0924-0136(01)01149-9
80. Singh R, Khamba JS (2006) Ultrasonic machining of titanium and its alloys: a review. J Mater Process Technol 173:125–135. doi:10.1016/j.jmatprotec.2005.10.027
81. Singh R, Khamba JS (2007) Taguchi technique for modeling material removal rate in ultrasonic machining of titanium. Mater Sci Eng A 460:365–369. doi:10.1016/j.msea.2007.01.093
82. Singh R, Khamba JS (2007) Investigation for ultrasonic machining of titanium and its alloys. J Mater Process Technol 183:363–367. doi:10.1016/j.jmatprotec.2006.10.026
83. Hasçalık A, Çaydaş U (2007) Electrical discharge machining of titanium alloy (Ti–6Al-4V). Appl Surf Sci 253:9007–9016. doi:10.1016/j.apsusc.2007.05.031
84. Gente A, Hoffmeister HW, Evans C (2001) Chip formation in machining Ti6Al4V at extremely high cutting speeds. CIRP Ann Manuf Technol 50:49–52. doi:10.1016/S0007-8506(07)62068-X
85. Komanduri R, Hou ZB (2002) On thermoplastic shear instability in the machining of a titanium alloy (Ti-6Al-4V). Metall Mater Trans A 33:2995–3010. doi:10.1007/s11661-002-0284-1

86. Lee D (1985) The effect of cutting speed on chip formation under orthogonal machining. J Eng Ind 107:55–63. doi:10.1115/1.3185966
87. Bäker M, Rösler J, Siemers C (2002) Finite element simulation of segmented chip formation of Ti6Al4V. J Manuf Sci Eng 124:485–488. doi:10.1115/1.1459469
88. Hua J, Shivpuri R (2004) Prediction of chip morphology and segmentation during the machining of titanium alloys. J Mater Process Technol 150:124–133. doi:10.1016/j.jmatprotec.2004.01.028
89. Sheikh-Ahmad J, Quarless V, Bailey J (2004) On the role of microcracks on flow instability in low speed machining of CP titanium. J Mach Sci Tech 8:415–430. doi:10.1081/MST-200039867
90. Shivpuri R, Hua J, Mittal P, Srivastava A, Lahoti G (2002) Microstructure-mechanics interactions in modeling chip segmentation during titanium machining. CIRP Ann Manuf Technol 51:71–74. doi:10.1016/S0007-8506(07)61468-1
91. Calamaz M, Nouari M, Géhin D, Girot F (2006) Damage modes of straight tungsten carbide in dry machining of titanium alloy TA6V. J de Phy IV 134:1265–1271. doi:10.1051/jp4:2006134192
92. Bayoumi A, Xie J (1995) Some metallurgical aspects of chip formation in cutting Ti-6wt.% Al-4wt.% V alloy. Mater Sci Eng A 190:173–180. doi:10.1016/0921-5093(94)09595-N
93. Sorby K, Tonnessen K (2006) High-pressure cooling of face-grooving operations in Ti6Al4 V. Proc Inst Mech Eng Part B J Eng Manuf 220:1621–1627. doi:10.1243/09544054JEM474
94. Machado A, Wallbank J, Pashby I, Ezugwu E (1998) Tool performance and chip control when machining Ti6A14V and Inconel 901 using high pressure coolant supply. J Mach Sci Tech 2:1–12. doi:10.1080/10940349808945655
95. Vyas A, Shaw M (1999) Mechanics of saw-tooth chip formation in metal cutting. J Manuf Sci Eng 121:163–172. doi:10.1115/1.2831200
96. Obikawa T, Usui E (1996) Computational machining of titanium alloy—finite element modeling and a few results. J Manuf Sci Eng 118:208–215. doi:10.1115/1.2831013
97. Komanduri R, Von Turkovich B (1981) New observations on the mechanism of chip formation when machining titanium alloys. J Wear 69:179–188. doi:10.1016/0043-1648(81)90242-8
98. Barry J, Byrne G, Lennon D (2001) Observations on chip formation and acoustic emission in machining Ti–6Al-4V alloy. Int J Mach Tools Manuf 41:1055–1070. doi:10.1016/S0890-6955(00)00096-1
99. Ulutan D, Erdem Alaca B, Lazoglu I (2007) Analytical modelling of residual stresses in machining. J Mater Process Technol 183:77–87. doi:10.1016/j.jmatprotec.2006.09.032
100. Hua J, Umbrello D, Shivpuri R (2006) Investigation of cutting conditions and cutting edge preparations for enhanced compressive subsurface residual stress in the hard turning of bearing steel. J Mater Process Technol 171:180–187. doi:10.1016/j.jmatprotec.2005.06.087
101. Niknam SA, Songmene V (2014) Milling burr formation, modeling and control: a review. Proc Inst Mech Eng Part B J Eng. Manuf (In press)
102. Niknam SA, Kamguem R, Songmene V (2012) Analysys and optimization of exit burr size and surface roughness in milling using desireability function. Proceedings of ASME 2012 international mechanical engineering congress & Expo. Paper No. IMECE2012-86201. Houston, Nov 9–15, 2012. doi:10.1115/IMECE2012-86201
103. Niknam SA, Tiabi A, Kamguem R, Zaghbani I, Songmene V (2011) Milling burr size estimation using acoustic emission and cutting forces. Proc. of ASME Int Mech Eng Cong Expo. Paper No. IMECE2011-63824. Denver, Nov 11–17. doi:10.1115/IMECE2011-63824
104. Niknam SA, Songmene V (2012) Statistical investigation on burrs thickness during milling of 6061-T6 aluminium alloy. Proc of CIRP 1st Int Conf on Virtual Machining Process Technology, Montreal, 28 May–1 June 2012
105. Niknam SA, Songmene V (2014) Analytical modelling of slot milling exit burr siz. Int J Adv Manuf Tech (In press). doi: 10.1007/s00170-014-5758-y

106. Zedan Y, Niknam SA, Djebara A, Songmene V (2012) Burr size minimization when drilling 6061-T6 aluminum alloy. Proc of ASME 2012 Int Mech Eng Cong & Expo. Paper No. IMECE2012-86412. Houston, Nov 9–15, 2012. doi: 10.1115/IMECE2012-86412

107. Niknam SA (2013) Burrs understanding, modeling and optimization during slot milling of aluminium alloys. Ph.D. Thesis, École de Technologie Superieure, Universite du Quebec

108. Niknam SA, Songmene V (2013) Modeling of burr thickness in milling of ductile materials. Int J Adv Manuf Tech 66:2029–2039. doi:10.1007/s00170-012-4479-3

109. Niknam SA, Songmene V (2013) Factors governing burr formation during high-speed slot milling of wrought aluminium alloys. Proc Inst Mech Eng Part B J Eng Manuf 227:1165–1179. doi:10.1177/0954405413484725

110. Niknam SA, Songmene V (2013) Simultaneous optimization of burrs size and surface finish when milling 6061-T6 aluminium alloy. Int J Precis Eng Manuf 14:1311–1320. doi:10.1007/s12541-013-0178-8

111. Niknam SA, Songmene V (2013) Experimental investigation and modeling of milling burrs. Proc of ASME 2013 Int Manu Sci and Eng Conf, Madison, June 10–14, 2013. doi:10.1115/MSEC2013-1176

112. Niknam SA, Wygowski W, Balazinski M, Songmene V (2014) Milling burr formation and aavoidance. In: Davim JP (ed) Machinability of advanced materials, ISTE Wiley, London, pp 57–94. doi:10.1002/9781118576854.ch2

113. Tsann-Rong L (2000) Experimental study of burr formation and tool chipping in the face milling of stainless steel. J Mater Process Technol 108:12–20. doi:10.1016/S0924-0136(00)00573-2

114. Barth C, Dollmeier R, Warnecke G (2001) Burr formation in grinding of hardened steel with conventional and superabrasive wheels. Proc of North Am Manuf Res Conf 2001,Gainsville, 22–25 May 2001

115. Lin TR (2002) Optimisation technique for face milling stainless steel with multiple performance characteristics. Int J Adv Manuf Tech 19:330–335. doi:10.1007/s001700200021

116. Davim JP, Gaitonde V, Karnik SR (2007) Integrating Taguchi principle with genetic algorithm to minimize burr size in drilling of AISI 316L stainless steel using an artificial neural network model. Proc Inst Mech Eng Part B J Eng Manuf 221:1695–1704. doi:10.1243/09544054JEM863

117. Gaitonde VN, Karnik SR, Achyutha BT, Siddeswarappa B (2008) Taguchi optimization in drilling of AISI 316L stainless steel to minimize burr size using multi-performance objective based on membership function. J Mater Process 202:374–379. doi:10.1016/j.jmatprotec.2007.08.013

118. Kim J, Min S, Dornfeld DA (2001) Optimization and control of drilling burr formation of AISI 304L and AISI 4118 based on drilling burr control charts. Int J Adv Manuf Tech 41:923–936. doi:10.1243/09544054JEM863

119. Karnik S, Gaitonde V, Davim JP (2007) Integrating Taguchi principle with genetic algorithm to minimize burr size in drilling of AISI 316L stainless steel using an artificial neural network model. Proc Inst Mech Eng Part B J Eng 221:1695–1704. doi: 10.1243/09544054JEM863

120. Aurich JC, Dornfeld D, Arrazola PJ, Franke V, Leitz L, Min S (2009) Burrs-analysis, control and removal. CIRP Ann Manuf Technol 58:519–542. doi:10.1016/j.cirp.2009.09.004

121. Dornfeld DA, Kim JS, Dechow H, Hewson J, Chen LJ (1999) Drilling burr formation in titanium alloy, Ti-6Al-4V. CIRP Ann Manuf Technol 48:73–76. doi:10.1016/S0007-8506(07)63134-5

122. Schueler G, Engmann J, Marx T, Haberland R, Aurich J (2010) Burr formation and surface characteristics in micro-end milling of titanium alloys. Proceedings of the CIRP international conference on Burrs, April 2–3, 2009, University of Kaiserslautern, Germany. doi:10.1007/978-3-642-00568-8_14

123. Konig W (1978) Applied research on the machinability of titanium and its alloys. Proc AGARD Conf in Advanced Fabrication Processes, Florence

124. Sun J, Guo Y (2008) A new multi-view approach to characterize 3D chip morphology and properties in end milling titanium Ti–6Al–4V. Int J Mach Tools Manuf 48:1486–1494. doi:10.1016/j.ijmachtools.2008.04.002
125. Ibrahim G, Haron C, Ghani J (2009) The effect of dry machining on surface integrity of titanium ally Ti-6Al-4V. J Appl Sci 9:121–127
126. Puerta Velasquez JD (2007) Etude des copeaux et de l'intégrité de surface en usinage à grande vitesse de l'alliage de titane TA6V
127. Velásquez J, Tidu A, Bolle B, Chevrier P, Fundenberger JJ (2010) Sub-surface and surface analysis of high speed machined Ti–6Al–4V alloy. Mater Sci Eng A 527:2572–2578. doi:10.1016/j.msea.2009.12.01
128. Mhamdi M, Boujelbene M, Bayraktar E, Zghal A (2012) Surface integrity of titanium alloy Ti-6Al-4V in ball end milling. Phys Procedia 25:355–362. doi:10.1016/j.phpro.2012.03.096
129. Moussaoui K, Mousseigne M, Senatore J, Lagarrigue P, Caumel Y (2013) Influence of milling on surface integrity of Ti-6Al-4V. Adv Mat Res 698:127–136. doi:10.4028/www.scientific.net/AMR.698.127
130. Sadeghi M, Haddad M, Tawakoli T, Emami M (2009) Minimal quantity lubrication-MQL in grinding of Ti–6Al-4V titanium alloy. Int J Adv Manuf Tech 44:487–500. doi:10.1007/s00170-008-1857-y
131. Shokrani A, Dhokia V, Newman ST (2012) Environmentally conscious machining of difficult-to-machine materials with regard to cutting fluids. Int J Mach Tools Manuf 57:83–101. doi:10.1016/j.ijmachtools.2012.02.002
132. Rahman M, Wang ZG, Wong YS (2006) A review on high-speed machining of titanium alloys. JSME JSME Int J Ser C 49:11–20
133. Nath KSG, Srivastava AK, Iverson J (2014) Study of droplet spray behavior of an atomization-based cutting fluid spray system for machining titanium alloys. J Manuf Sci. doi:10.1115/1.4025504
134. Hafenbraedl D, Malkin S (2001) Technology environmentaly correct for intern cylindrical grinding, Mach Metals Magaz 426:40–55
135. Su Y, He N, Li L, Li XL (2006) An experimental investigation of effects of cooling/lubrication conditions on tool wear in high-speed end milling of Ti-6Al-4V. J Wear 261:760–766. doi:10.1016/j.wear.2006.01.013
136. Heisel U, Lutz D, Wassmer R, Walter U (1998) The minimum quantity lubricant technique and its application in the cutting process. Mach Metals Magaz 386:22–38
137. Hong SY, Zhao Z (1999) Thermal aspects, material considerations and cooling strategies in cryogenic machining. J Clean Products Process 1:107–116. doi:10.1007/s100980050016
138. O'sullivan D, Cotterell M (2001) Temperature measurement in single point turning. J Mater Process Technol 118:301–308. doi:10.1016/S0924-0136(01)00853-6
139. Niknam SA, Kouam J, Songmene V (2014) Experimental investigation on surface finish, burr formation and particles emission during slot milling of 6061-T6 aluminum alloy. Proceedings of ASME 2014 international mechanical engineering congress & Expo IMECE2014, Nov 14–20, 2014, Montreal
140. Dhar N, Islam M, Islam S, Mithu M (2006) The influence of minimum quantity of lubrication (MQL) on cutting temperature, chip and dimensional accuracy in turning AISI-1040 steel. J Mater Process Technol 171:93–99. doi:10.1016/j.jmatprotec.2005.06.047
141. Yue Y, Sun J, Gunter K, Michalek D, Sutherland J (2004) Character and behavior of mist generated by application of cutting fluid to a rotating cylindrical workpiece. Part 1: model development. J Manuf Sci Eng 126:417–425. doi:10.1115/1.1765150
142. Chen D, Sarumi M, Al-Hassani S (1998) Computational mean particle erosion model. J Wear 214:64–73. doi:10.1016/S0043-1648(97)00210-X
143. Chen Z, Stephenson DA, Wong K, Li W, Liang SY (2001) Cutting fluid aerosol generation due to spin-off in turning operation: analysis for environmentally conscious machining. J Manuf Sci Eng 123:506–512. doi:10.1115/1.1367268

144. Chen Z, Atmadi A, Stephenson DA, Liang SY, Patri KV (2000) Analysis of cutting fluid aerosol generation for environmentally responsible machining. CIRP Ann Manuf 49:53–56. doi:10.1016/S0007-8506(07)62894-7

145. Chen Z, Liang SY, Yamaguchi H (2002) Predictive modeling of cutting fluid aerosol generation in cylindrical grinding. Soc Manuf Eng (SME)196:1–8

146. Atmadi A, Stephenson A, Liang S (2001) Cutting fluid aerosol from splash in turning: analysis for environmentally conscious machining. Int J Adv Manuf Tech 17:238–243. doi:10.1007/s001700170175

147. Rossmoore H, Rossmoore L (1991) Effect of microbial growth products on biocide activity in metalworking fluids. Int J Biodeterior Biodegrad 27:145–156. doi:10.1016/0265-3036(91)90006-D

148. Sondossi M, Rossmoore H, Williams R (1989) Relative formaldehyde resistance among bacterial survivors of biocide-treated metalworking fluid. Int J Biodeterior Biodegrad 25:423–437. doi:10.1016/0265-3036(89)90068-7

149. Sutherland J, Kulur V, King N, Von Turkovich B (2000) An experimental investigation of air quality in wet and dry turning. CIRP Ann Manuf Technol 49:61–64. doi:10.1016/S0007-8506(07)62896-0

150. Zaghbani I, Songmene V, Khettabi R (2009) Fine and ultrafine particle characterization and modeling in high-speed milling of 6061-T6 aluminum alloy. J Mater Eng Perform 18:38–49. doi:10.1007/s11665-008-9265-x

151. Songmene V, Masounave J, Balout B (2008) Clean machining: experimental investigation on dust formation part II: influence of machining parameters and chip formation, part II. Int J Environ Conscious Des Manuf(ECDM) 14:17–33

152. Khettabi R, Songmene V, Masounave J (2007) Effect of tool lead angle and chip formation mode on dust emission in dry cutting. J Mater Process Technol 194:100–109. doi:10.1016/j.jmatprotec.2007.04.005

153. Khettabi R, Songmene V, Masounave J (2010) Effects of speeds, materials, and tool rake angles on metallic particle emission during orthogonal cutting. J Mater Eng Perform 19:767–775. doi:10.1007/s11665-009-9551-2

154. Khettabi R, Songmene V, Zaghbani I, Masounave J (2010) Modeling of particle emission during dry orthogonal cutting. J Mater Eng Perform 1919:776–789. doi:10.1007/s11665-009-9538-z

155. Khettabi R, Songmene V, Masounave J, Zaghbani I (2008) Understanding the formation of nano and micro particles during metal cutting. Int J Syst Signal control Eng appl 1:203–210

156. Khettabi R, Songmene V (2009) Particle emission during orthogonal and oblique cutting. Int J Adv Mach and Form Oper 1:1–9

157. Ezugwu EO (2005) Key improvements in the machining of difficult-to-cut aerospace superalloys. Int J Mach Tools Manuf 45:1353–1367. doi:10.1016/j.ijmachtools.2005.02.003

158. Khettabi R, Fatmi L, Masounave J, Songmene V (2013) On the micro and nanoparticle emission during machining of titanium and aluminum alloys, CIRP. J Manufact Sci Technol 6:175–180. http://dx.doi.org/10.1016/j.cirpj.2013.04.001

Chapter 2
Cutting Tool Materials and Tool Wear

Ali Hosseini and Hossam A. Kishawy

Abstract The chip formation in machining operations is commonly accomplished by a combination of several elements working together to complete the job. Among these components, cutting tool is the key element that serves in the front line of cutting action. Cutting action becomes a challenge when it comes to machining difficult-to-cut materials. Titanium and its alloys are among the most difficult-to-cut materials which are widely used in diverse industrial sectors. This chapter aims to provide a historical background and application of different cutting tools in machining industry with a main focus on the applicable cutting tools in machining titanium and titanium alloys. Selection of appropriate tool material for a certain application is directly influenced by the characteristics of material to be machined. In this context, a brief overview of the metallurgy of titanium and its alloys is also presented. Recent progresses in tool materials, appropriate tools for cutting titanium alloys, and their dominant wear mechanisms will also be covered in this chapter.

2.1 Introduction

Nowadays, aerospace, power generation, oil and gas, marine, and medical industries are among rapidly developing business which plays an important role in almost every aspects of the human's life as well as the global economy. Due to their inherent nature, the majority of mechanical parts employed in these industries are usually used in severe climate conditions. Structural parts of aircrafts' fuselage, components of jet engines, blades of compressors and turbines, combustion

A. Hosseini · H. A. Kishawy (✉)
Machining Research Laboratory (MRL), Faculty of Engineering and Applied Science,
University of Ontario Institute of Technology (UOIT), Oshawa, ON L1H 7K4, Canada
e-mail: hossam.kishawy@uoit.ca

A. Hosseini
e-mail: sayyedali.hosseini@uoit.ca

J. P. Davim (ed.), *Machining of Titanium Alloys*,
Materials Forming, Machining and Tribology, DOI: 10.1007/978-3-662-43902-9_2,
© Springer-Verlag Berlin Heidelberg 2014

chambers, and exhaust nozzles are among several examples of such components. Typical material to be used in these applications must possess wide range of desirable properties which include but not limited to low density and high specific strength to weight ratio especially at elevated temperature, resistance to corrosion, and chemical inertness.

Titanium is a favourite choice and one of the most extensively used material for the above-mentioned applications as its specific strength to weight ratio especially at high temperatures is higher than its steel and aluminium counterparts, which makes it the material of choice for aerospace and power generation industries. In addition, titanium and its alloys exhibit remarkable resistance to corrosion and are capable of sustaining in marine environments where other architectural materials such as metals demonstrate limited lifecycle. Such corrosive environments are also very common in oil and gas industries. Furthermore, titanium shows superior elasticity which is a desirable characteristic for flexible parts when cracks or disintegration must be suppressed. Showing promising non-magnetic characteristics, titanium has been also used in computer industry as a substrate for hard disk drives which promotes data storage process by avoiding any electromagnetic interference. Chemical inertness or bio-adaptability is another desirable characteristic that titanium possesses which makes it an ideal candidate for medical applications such as implants.

Despite several advantages offered by titanium in comparison to the other commonly used materials in industries, many manufacturing challenges arise when it comes to machining titanium and it alloys. These challenges are mainly originated from mechanical, thermal and chemical characteristics of titanium. These characteristics include low modulus of elasticity, poor thermal conductivity, chemical reactivity at high temperatures, and finally hardening characteristics. As a result of these features, titanium is being classified as a hard-to-cut or difficult-to-cut material. Due to these inherent characteristics, cutting tools for machining titanium and its alloys must be wisely selected to mitigate these machining challenges.

2.2 What is Titanium?

Titanium is a silver colored shiny metal with the atomic number of 22 and the chemical symbol of Ti which was initially discovered in 1791 by William Gregor, an English chemist and mineralogist. Martin Heinrich Klaproth, a German chemist, named this newly discovered element titanium for the Titans of Greek mythology [1]. Titanium constitutes 0.565 % of the earth crust and is the 9th plentiful element and also the 4th plentiful structural metal in the earth crust after aluminum (8.23 %), iron (5.63 %), and magnesium (2.33 %) [2]. Only aluminum (8.23 %), iron (5.63 %), and magnesium (2.33 %) can be found more than titanium in earth's crust. The magnitude of obtainable titanium is more than zinc, copper, nickel, tin, lead, and chromium put together [2].

Titanium can be found in two major commercial minerals namely ilmenite and rutile [1]. It can be also found broadly all around the world in natural waters, animals and plants' bodies, sands, and rocks. Ilmenite is a crystalline iron titanium oxide ($FeTiO_3$) which is steel-gray or iron-black. In contrast, rutile is a mineral composed primarily of titanium dioxide (TiO_2) with blood red or brownish color.

It can come into question that despite its relatively widespread presence in nature, why titanium is very expensive and is not as widely used as other conventional engineering metals such as aluminium [3]. Titanium is an expensive material because its extraction process is tremendously costly and labor intensive. The Kroll method which is currently being used to extract and refine the titanium consists of several steps that must be performed for each batch of ore at high temperature [2].

Difficulties toward extraction of titanium can be summarized as follows [4]:

- Reducing agents like carbon cannot be used to reduce the ore because it forms titanium carbide (TiC), if carbon and titanium are heated together. The resultant product will not be pure metal and will be very brittle due to the presence of the carbide.
- The alternative options are either sodium or magnesium as reducing agents which are also expensive to extract from their ores.
- In addition, the titanium oxide (TiO_2) must be first converted to the titanium chloride ($TiCl_4$) to make it able to react with sodium or magnesium. As a result, the chlorine cost is also imposed to the cost of titanium in addition to the other cost such as energy cost.
- The titanium chloride must be handled carefully to prevent any contact with water because it aggressively reacts with water.
- The presence of oxygen or nitrogen makes the titanium brittle; hence, the reduction process of titanium must be performed in an inert argon atmosphere rather than in the air.
- Despite production of iron in the blast furnaces which is a continuous and efficient process, titanium is extracted from its ore in a batch process at which titanium chloride is heated with sodium or magnesium to produce titanium. This process generates some waste products that must be separated to achieve pure titanium. The whole process should be entirely set up again for a new batch which makes the process very slow and inefficient.

Four different approaches are currently used to extract the titanium from its ore. These approaches include Kroll, Hunter, Cambridge, and Armstrong processes [2] among them the Kroll method has been the prevailing commercial process for production of titanium since the 1940s [3, 5]. The following chemical equations show the basic concept of the Kroll method [3, 6].

$$Tio_2 + 2Cl_2 + C \rightarrow TiCl_4 + CO_2 \qquad (2.1)$$

$$TiCl_4 + 2Mg \rightarrow Ti + 2MgCl_2 \qquad (2.2)$$

Reaction (2.1) produces an oxygen free tetrachloride from titanium dioxide and reaction (2.2) forms the titanium sponge. The titanium sponge is then further processed depending on the final product applications. The detailed description of the Kroll method can be found in [2, 3, 6].

2.3 Metallurgy of Titanium and Its Alloys

Titanium has two types of crystal structure namely Alpha (α) and Betta (β) [7]. In temperatures below 882 °C, titanium can be found in hexagonal closely packed (hcp) α phase crystal structure. While temperature goes beyond 882 °C, the α phase undergoes an allotropic transformation to a body-centered cubic (bcc) β phase. This phase remains stable up to the melting point of titanium (1,668 °C) [8, 9]; however, adding certain elements may alter the transformation temperature [9]. Figure 2.1 shows the two allotropic forms of titanium.

Adding Aluminum (Al), Gallium (Ga), Oxygen (O), Nitrogen (N), and Carbon (C) raises the transformation temperature. These elements stabilize the α phase and they are known as α stabilizer. In contrast, applying elements such as Vanadium (V), Molybdenum (Mo), Niobium (Nb), Iron (Fe), Chromium (Cr), Nickel (Ni), Manganese (Mn), and Cobalt (Co) lowers the transformation temperature. Similar to α stabilizer, these elements are called β stabilizer [8, 9]. One of the most commercially used element in titanium alloys is aluminium which is among the α stabilizers [8].

An impure material which can be a mixture of either pure or relatively pure chemical elements with an additive metallic material is called alloy. The additive materials are normally called elements while the primary metal which the elements are added to is usually called base metal. The alloy normally preserves the positive features of a base metal while adding some additional valuable benefits. The mechanical properties of alloy might be quite different from those of base metal as well as its individual constituents.

Although pure (unalloyed) titanium shows acceptable corrosion resistance, it is not being used in its pure state. Titanium is commonly alloyed with small amounts of some other elements such as Aluminium (Al) and Vanadium (V) to promote mechanical properties [9]. Titanium alloys can be divided into several categories based on the alloying condition and possible additive elements that can be added to the microstructure of titanium [8].

2.3.1 Alpha (α) Alloys

The alpha (α) alloys are single phase titanium alloys which consist of α stabilizer or some other neutral alloying elements [7, 8]. These titanium alloys maintain their tensile strength up to 300 °C and also exhibit exceptional creep stability. These

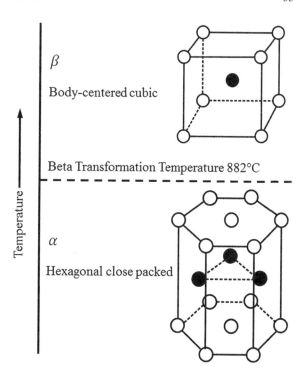

Fig. 2.1 The two allotropic forms of titanium [9]

titanium alloys are not heat treatable and their microstructural properties cannot be modified by heat treatment. As these alloys lose their tensile strength in temperatures above 300 °C, their primary applications are cryogenic applications or where great resistance to corrosion is compulsory. One of the frequently used alpha (α) alloys is Ti5-$2\frac{1}{2}$ (Ti-5Al-$2\frac{1}{2}$Sn) [7].

2.3.2 Near Alpha Alloys

It has been proven [8] that workability and strength of titanium alloys can be improved by adding a small portion (1–2 %) of β stabilizers. The near alpha (α) titanium alloys are highly α-stabilised and include small quantities of β-stabilizing elements. These alloys are described by their α phase microstructure containing small quantities of β phase. Due to their high similarity to α phase alloys, near alpha (α) alloys are capable of working at elevated temperatures between 400 and 520 °C [9]. Hence, near alpha (α) titanium alloys are primary candidates for aerospace applications, especially components of jet engines where the parts are being utilized at elevated temperatures. Ti 8-1-1 (Ti-8Al-1Mo-1V) and IMI 685 (Ti-6Al-5Zr-0.5Mo-0.25Si) are some examples of this family of titanium alloys [7].

2.3.3 Alpha-Beta (α + β) Alloys

If the amount of beta stabilizer added to the titanium is larger (4–6 %) than that in near alpha (α) alloys (1–2 %) a new category of titanium alloys called α + β alloys will be generated [8]. Different combinations of microstructures and consequently mechanical properties can be developed by heat treating α + β alloys. The heat treatment improves the strength and makes this category of titanium alloys a principal choice for elevated temperature's (350–400 °C) application. IMI 550 (Ti-4Al-2Sn-4Mo-0.5Si) and specially Ti 6-4 (Ti-6Al-4V) which is one of the most commonly used titanium alloys in industry belong to this group [7].

2.3.4 Metastable Beta (β) Alloys

By further increasing the amount of β stabilizers (10–15 %), β phase is retained in a metastable state at room temperature [8]. The metastable β alloys include a small portion of α stabilizers to increase the strength. Demonstrating high toughness, high strength, great hardenability, and forgeability over a wide range of temperatures, this family of titanium alloys is a potential candidate for structural parts in aerospace applications.

2.3.5 Beta (β) Alloys

A titanium alloys which contain a large amount (30 %) of β stabilizers is called beta β alloy. Due to high density and also poor ductility, this family of titanium alloys has some particular applications specially where burn-resistance and corrosion-resistance is required [8].

2.3.6 Titanium Aluminides

Titanium aluminide (TiAl) is an intermetallic chemical compound comprises three main intermetallic compounds: gamma TiAl (Ti3Al (γ)), alpha 2-Ti3Al (Ti3Al (α2)) and TiAl3 [10].

The term intermetallic or intermetallic compound is generally refers to solid-state phases containing metals. Although both alloys and intermetallic compound contain more than one element, they cannot be categorized under a same classification due to some differences. Alloy is normally referred to a solid solution of a base metal and some various elements which has metallic properties while intermetallic materials are chemical mixtures of two or more metals. The crystal structure of intermetallic materials are different from the crystal structure of the metals it's made from.

To achieve some superior properties such as excellent heat and oxidation resistance, pure titanium can be alloyed by titanium aluminide. However, despite their great characteristics, aluminide-based titanium alloys exhibit poor ductility and low fracture toughness [8].

2.4 Titanium as a Hard-to-Cut Material

In spite of having several available academic resources and research papers on machining and machinability of materials, defining a certain border between hard materials and hard to cut materials is still a challenging task. Hence, the differences between hard material and hard-to-cut material must be clarified prior to any discussion about machining and machinability of titanium.

Among the mechanical properties, strength and hardness have the highest impact on the machining performance or simply ease of machining for certain material [11]. It may be concluded that increasing the material strength leads to larger cutting forces and higher temperatures which make the material more difficult to cut. However, machining tests shows that the materials with higher strength or hardness do not necessarily require larger cutting forces in machining. It has been shown [12, 13] that machining of medium carbon steel AISI 1045 (ultimate tensile strength $\sigma_R = 655$ MPa, yield strength $\sigma_{y0.2} = 375$ MPa) requires lower cutting force and lower cutting power which results in lower cutting temperature, lower residual stresses, and greater tool life in comparison to those acquired in the machining of stainless steel AISI 316L ($\sigma_R = 517$ MPa, $\sigma_{y0.2} = 218$ MPa) [14]. Higher hardness of work material is another factor that accelerates the tool wear and decrease the tool life which is one of the indicators of machinability [11]. Table 2.1 shows the approximate values of hardness and typical machinability ratings for some work materials.

As can be seen in the Table 2.1, comprising low hardness does not necessarily mean that the material with lower hardness can be easily machined. Very low hardness negatively affects the machining performance. For example, low carbon steel which has relatively low hardness, is usually classified under the material with low machinability due to its high ductility. High ductility results in the poor surface finish due to tearing of the metal during chip formation [11].

For instance, the machinability rating for steels AISI 8620 and AISI 8630 with brinell hardness ranging from 190 to 200 is 0.6; while, for plain titanium with brinell hardness of 160, this rating dramatically drops to 0.3 (see Table 2.1). Taking above-mentioned criteria into consideration, it can be concluded that although machinability of a certain material can be dramatically affected by hardness; however, hardness is not a unique performance measure for ease or difficulty of machining.

Table 2.1 Approximate values of Brinell hardness and typical machinability rating for selected work materials [11]

Work material	Brinell hardness	Machinability rating
Base steel: B1112	180–220	1.00
Low carbon steel: C1008, C1010, C1015	130–170	0.50
Medium carbon steel: C1020, C1025, C1030	140–210	0.65
High carbon steel: C1040, C1045, C1050	180–230	0.55
Alloy steels 24		
1320, 1330, 3130, 3140	170–230	0.55
4130	180–200	0.65
4140	190–210	0.55
4340	200–230	0.45
4340 (casting)	250–300	0.25
6120, 6130, 6140	180–230	0.50
8620, 8630	190–200	0.60
B1113	170–220	1.35
Free machining steels	160–220	1.50
Stainless steel		
301, 302	170–190	0.50
304	160–170	0.40
316, 317	190–200	0.35
403	190–210	0.55
416	190–210	0.90
Tool steel (unhardened)	200–250	0.30
Cast iron		
Soft	60	0.70
Medium hardness	200	0.55
Hard	230	0.40
Super alloys		
Inconel	240–260	0.30
Inconel X	350–370	0.15
Waspalloy	250–280	0.12
Titanium		
Plain	160	0.30
Alloys	220–280	0.20
Aluminum		
2-S, 11-S, 17-S	Soft	5.00
Aluminum alloys (soft)	Soft	2.00
Aluminum alloys (hard)	Hard	1.25
Copper	Soft	0.60
Brass	Soft	2.00
Bronze	Soft	0.65

The term machinability is commonly used to describe the ease or difficulties of machining for a certain work material. The machining performance is governed by several contributing factors in addition to mechanical properties of work material. Machining processes, cutting tools, and cutting conditions are among the important factors that affect the machinability of certain material on the top of its mechanical properties.

Generally speaking, although hard materials are generally hard to cut, but every hard to cut material is not necessarily hard. In the other word, hard to cut materials or materials with low machinability are not necessarily very hard materials. It may come into question that why titanium is a hard-to-cut material. This question will be answered in the next section.

Historically, titanium has been always considered as a material that is difficult to machine. As a result of the widely acceptance and the emerging application of titanium in many industries, valuable experiences accompanied by a broad base of knowledge have been acquired regarding machining titanium and its alloys.

It has been determined that when it comes to machining titanium and its alloys, the majority of tool materials that show great performance in machining other materials, exhibit moderate to poor performance. Difficulties in machining titanium alloys are caused by a combination of the following features.

2.4.1 Poor Thermal Conductivity

During the cutting process, energy is consumed to form the chip by plastically deforming the workpiece body or to overcome friction. Almost all of this energy is then converted into heat and consequently increases the temperature in the cutting zone. The main heat sources during cutting operations are as follows (see Fig. 2.2):

- Primary shear zone at which the heat is mainly generated by plastic deformation of workpiece due to shearing.
- Secondary shear zone at which the heat is generated by a combination of shearing and friction on the tool rake face.
- Tertiary shear zone at which the heat is produced due to friction between newly machined and the flank face of the cutting tool.

Although the heat generated at the cutting zone softens the workpiece material and facilitate easier cutting, it is generally considered as an undesirable phenomenon that must be prohibited or kept minimized. The heat generated during machining is primarily dissipated by the discarded chip. A quite smaller portion of the heat is also dissipated by means of workpiece and cutting tool.

Due its poor thermal conductivity (about 15 W/m °C), the heat generated during machining titanium and its alloys is not easily dissipated from the cutting zone [7]; therefore, a vast amount of heat is trapped on or near the cutting zone which intensifies the temperature. It has been observed [7, 15, 16] that depending on the tool material, up to 50 % of the heat generated during machining can be transferred to the cutting tool. However, this magnitude may reach 80 % during machining titanium and its alloys. Figure 2.3 compares the distribution of thermal load when machining titanium alloy Ti-6Al-4V and Steel CK45.

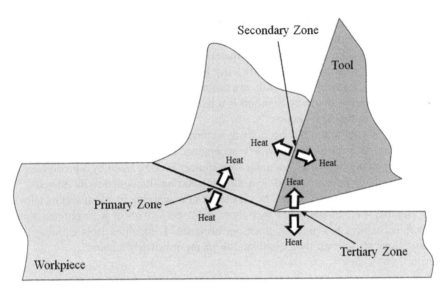

Fig. 2.2 Cutting zones and their corresponding generated heat

The concentrated heat at the cutting zone when machining titanium and its alloys sometimes reaches 1,100 °C [17]. The elevated temperature near the cutting zone where tool and workpiece are in touch can rapidly deteriorate the cutting edge and make it dull. Continuing machining using a tool with dull cutting edge may generates more heat and ceases the tool life.

2.4.2 Low Modulus of Elasticity and the Consequent Springback

Titanium has superior elasticity which makes it an ideal candidate for those applications where flexibility without the possibility of cracks or disintegration is desired. However, titanium's elasticity imposes another barrier toward machining. In comparison to the other metals, titanium's modulus of elasticity is relatively lower. Low modulus of elasticity results in relatively higher strain and consequently more deformation under a certain magnitude of force. In the other word, low modulus of elasticity makes titanium more bouncy [18]. During machining operations, when the tool touches the workpiece and cutting force is applied, titanium's elasticity makes the workpiece spring away from the tool, which causes cutting edge being rubbed against the workpiece surface rather than performing cutting action. Rubbing rather than cutting increases the friction and consequently further raises the temperature at the cutting zone. Rubbing also destroys the surface quality and dimensional accuracy.

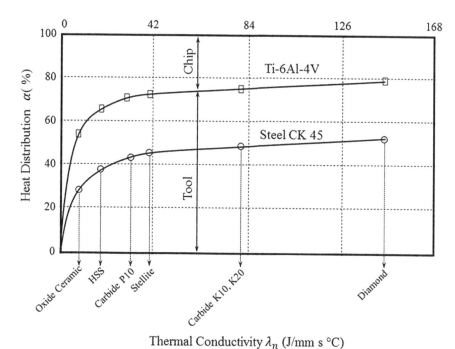

Fig. 2.3 Distribution of thermal load when machining titanium and steel [7, 16]

2.4.3 Chemical Reactivity

Despite its chemical inertness at room temperatures which makes it one of the best options for medical implants, titanium becomes highly reactive tools when the temperature goes beyond 500 °C. When the temperature increases, chemical reaction occurs between titanium and cutting tools which quickly obliterate the tool [7, 9]. As a result, the majority of currently available cutting tools, even the hardest ones, are not appropriate for machining titanium and its alloys due to chemical affinity which deteriorate the cutting tool by initiating chemical wear.

2.4.4 Hardening due to Diffusion and Plastic Deformation

It has been previously established that due to its poor thermal conductivity, the heat generated during machining titanium cannot be freely dissipated from the cutting zone. This localized heat is capable of raising temperature up to 1,100 °C.

When the temperature at the cutting zone reaches the range of 600–700 °C or exceeds this limit, oxygen and nitrogen molecules in the air are being diffused into the titanium workpiece and harden its surface layer [19].

In addition, in such an elevated temperature, most metals undergo thermal softening and lose their strength. Thermal softening can be considered as a desired phenomenon provided that it does not adversely affect the mechanical characteristics of workpiece as well as cutting tool. Thermal softening lowers the forces and energy required to perform machining operation.

However, when it comes to machining titanium and its alloys, the story will be different. Titanium preserves its superior strength at elevated temperatures [20]; hence, one of the main advantages of titanium which makes it an ideal choice for many applications turns into one of the principal challenges during machining. Maintaining its mechanical strength at high temperatures, comparatively higher cutting power is required to plastically deform titanium and produce chip. High cutting power and its consequent plastic deformation significantly hardens the machined surface. This is what usually referred to as strain-hardening or work-hardening.

It must be mentioned that, both diffusion and plastic deformation are influential in hardening titanium during machining operation; however plastic deformation is the predominant factor [9, 15].

2.4.5 Mechanism of Chip Formation

When machining titanium and its alloys, the chip is either formed by the propagation of crack from the exterior surface of the chip or development of adiabatic shear band which is primarily originated by the localized shear deformation [21–24]. In case of adiabatic shear, the machining is dominated by thermal softening rather than strain hardening [21, 22]. Localization of Shear leads to significant periodic variation of machining forces and subsequently chatters vibration [25]. Cyclic variation of machining forces is not a desirable phenomenon as it imposes fatigue to the cutting tool or may cause chipping or breakage of cutting tool.

It can be concluded from above-mentioned items that, titanium and its alloys comprise some unique mechanical and metallurgical characteristics that make them comparatively harder to cut than their other counterparts with equivalent hardness.

In order to achieve an acceptable metal removal rate (MRR) at reasonable cost, appropriate tools, machining conditions, and processing sequence must be selected properly.

Considering all above-mentioned challenges that can be faced during machining titanium and its alloys, a successful cutting tool must [26, 27]:

- Capable of maintaining hardness at high cutting temperatures due to the presence of extreme heat at the cutting zone (hot hardness).
- Possess good chipping resistance which is principally attributed to the formation of segmented chips.

- Demonstrate toughness and fatigue resistance to withstand against the cyclic machining forces during the formation of the segmented chips.
- Have low chemical affinity with titanium to minimize the possibility of reaction between tool and workpiece.
- Have high compressive strength.
- Have excellent thermal conductivity to scatter the heat generated during machining away from the cutting zone.

2.5 Cutting Tools: Historical Background and Chronological Advances

Cutting tools are frequently used in our every day's life. The regularly used cutting tools can be in the form of knives, razor blades, lawnmowers or more industrial tools in wood or metal working. Despite their widespread applications in modern lives, not too many questions have been raised about the origin and history of these tools. In the context of metal cutting or in general machining, a cutting tool is an instrument by means of which the metal is being removed from the workpiece body. In order to achieve successful cutting, cutting tools must be mechanically harder than the material to be machined.

Although cutting tools in their general form have been used by human beings for centuries; their modern history began during the industrial revolution in the nineteenth century. However, in the absence of systematic tool production before the twentieth century, the majority of the tools were prepared by their end users at local machine shops. As a result, having a combined knowledge of physics, chemistry, heat treatment, and also blacksmithing was among the necessary requirement for being a successful machinist. Before the twentieth century, cutting tools were mostly produced using carbon tool steels. These types of steel comprise high carbon content and can be successfully hardened.

One of the earliest-reported advances in cutting tool history was made in 1868 by Robert Forester Mushet, a British metallurgist, who discovered that hardness of steel and consequently tool life can be improved by adding tungsten [28] which has the second highest melting point of all elements after carbon and the highest melting point of all non-alloyed metals. Mushet steel is considered to be the first tool steel [28] which was later led to the discovery of high speed steels [29].

The emerging need for cutting tool material capable of enduring higher cutting speeds and resulting high temperatures led to a significant development which was made by American mechanical engineer Frederick Winslow Taylor during late 19th and early 20th century. He studied the cutting tools and their corresponding performances and proposed a novel tool life equation which is, in its augmented form, still one of the most widely used equations in metal cutting science and machining industry. Taylor also discovered that more durable steel, which is able to maintain its hardness at high temperature, can be achieved if it is being heated close to its melting point. This type of hardened steel can be assumed to be the first generation of high speed steel (HSS) tools.

The introduction of HSS tool increased the practicable cutting speed four times in comparison to the previously used carbon steels [30]. In comparison to carbon steel tools, HSS tools owe their superiority to the alloying elements. The alloying elements make the steel harder and more heat resistant [31].

HSS tools can be divided into almost thirty different grades; while, all of these grades can be categorized in three principal classifications: molybdenum based grades (M series), tungsten based grades (T series), and molybdenum-cobalt based grades. Among these grades, M and T series are the most commonly used HSS tools in industry.

Performance of HSS tools can be further increased by application of coating. Different types of coating can be used to cover the surface of HSS tools; among them, titanium nitride is the most effective one that increase allowable cutting speed as well as tool life. Titanium nitride can be deposited on the HSS surface by means of physical vapour deposition (PVD) techniques.

Another material that was introduced to the cutting tool industry in early 20th was Stellite which is a non-magnetic, wear and corrosion resistant alloy of cobalt and chromium. The progressing trend in development of more advanced and durable cutting tool were further continued by the introduction of cemented carbide around the 1920s and ceramic inserts after the Second World War.

By the evolution of science and technology, traditional HSS and cobalt steel cutting tools were far outnumbered by new cutting tools made from carbides and ceramics. These tool materials are even now among the most widely used cutting tools for mass production of industrial parts.

Developed around 1930s, carbide tools comprise high modulus of elasticity, high thermal conductivity, and ultimately high hardness over a wide range of temperatures. Carbide tools, either uncoated or coated, are capable of reaching cutting speed of three to five times higher than their HSS counterparts [30]. Tungsten carbide (WC), titanium carbide (TiC), tantalum carbide (TaC_x), and niobium carbide (NbC and Nb_2C) are the most recognized hard carbides that can be used toward making carbide tools in industry. A typical carbide tool comprises these carbide particles bounded together in a matrix of cobalt using sintering process [32]. Among the carbide tools, tungsten carbide with 6 % cobalt binder was initially introduced to the industry in Germany in 1926 [32].

The mechanical characteristics and performance of carbide tools are greatly affected by the type of carbide. For instance, increasing the tungsten content will increase the wear resistance, but adversely affect the tool strength. Cobalt content also affects the mechanical properties of carbide tools. Increasing the cobalt content improves the toughness of the tool; however, it reduces the strength, hardness, and consequently wear resistance [32]. In comparison to the tungsten carbide, titanium carbide shows relatively higher wear resistance and lower toughness.

Ceramic cutting tools are another widely used category of cutting tools which were introduced to industry in the middle of the twentieth century. Ceramic tools mainly contain aluminum oxide (Al_2O_3), silicon nitride (Si_3N_4), and sialon (a combination of silicon (Si), aluminum (Al), oxygen (O), and nitrogen (N)) grains sintered together under high temperature (1,700 °C) and pressure (more than 25 MPA) [32, 33]. In comparison to the majority of cutting tools that demonstrate a rapid softening rate at elevated temperatures, ceramic tools exhibit much slower rate and capable of retaining their hardness in such conditions. Despite their hot hardness, ceramic tools suffer from lack of toughness; as a result, any type of shocks or impact during machining must be avoided to prevent chipping or breakage.

New materials with superior characteristics had been continually introduced to the market during the twentieth century. These new materials were mostly utilized in such applications where high performance reliability was required during the service life. To cope with the rapid growth of industries and their corresponding necessities, more effective machining should have been applied. To achieve this ultimate objective, cubic boron nitride tools, polycrystalline cubic boron nitride tools, and polycrystalline diamond tools were introduced to the industrial market.

Exhibiting hardness of up to 4,500 HV, cubic boron nitride tools are the second hardest ever existing tools after diamond with hardness of more than 9,000 HV. CBN is a polymorph boron-nitride-based material which was introduced to the industry in 1957. This family of cutting tools owes its superior mechanical properties to their crystalline structure and its covalent link [33]. Cubic boron nitride is produced by exposing hexagonal boron nitride to elevated temperatures of up to 1,500 °C and extreme pressure 8 GPa [33]. CBN offers excellent high hot hardness at up to 1,500 °C and even sometimes higher up to 2,000 °C. Due to their extreme hardness, CBN tools show great wear resistance; however, they suffer from lack of toughness. Polycrystalline cubic boron nitride tools are produced by sintering cubic boron nitride crystals with a binder under high temperature, and high pressure.

Diamond is the hardest ever existing material that combines some desired properties such as extreme hardness, highest thermal conductivity at room temperature and low coefficient of surface friction all together [34]. Despite their early applications as a cutting tool, due to their brittleness, single crystal diamonds need to be used at the correct crystal orientation to achieve optimum performance and prevent fracture. To incorporate the superior characteristics of diamond while eliminating their weakness, single crystal diamond tools have been substituted by polycrystalline diamond (PCD) tools. PCD is a composite of diamond particles sintered together with a metallic binder under high temperature and pressure. Due to its extreme hardness, PCD tools demonstrate wear resistance almost 500 times greater than those of tungsten carbide [32]. Due to their high hardness, similar to CBN and PCBN tools, PCD tools are extremely fragile and exhibit low toughness. They also chemically react with iron which makes them an inappropriate option for machining steels.

Generally speaking, different types of cutting tool materials with divers characteristics are now being used in industry. Although these tool materials are coming from different origins with various properties, some performance measures are required to compare them and make a judgement about the applicability of a certain tool for a particular application. These performance measures are hardness, toughness and wear resistance [32].

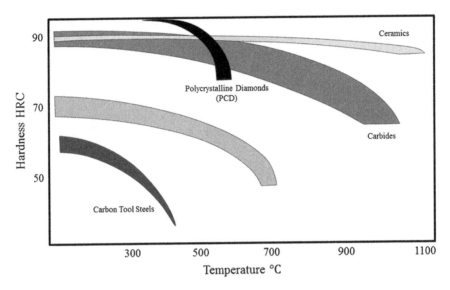

Fig. 2.4 Comparison of the capability of the major families of cutting tools as a function of temperature

Hardness is generally considered as the strength of intermolecular bonds in maintaining their shape without any permanent deformation. It can be also interpreted as the ability of a material in localizing deformation. In the context of cutting tools, hardness is defined as the ability to penetrate into the softer materials (workpiece). Hardness is also a performance measure which describes the capability of tool material in resisting against the permanent changes in shape and geometry during machining. This characteristic becomes more important when the cutting tool is exposed to the extreme heat generated during cutting operation. In this case, a successful cutting is the one with hot hardness which is capable of maintaining its hardness at high temperature [32]. Figure 2.4 compares the capability of the major families of cutting tools under temperature variation.

However, it must be pointed out that extreme hardness is not necessarily a desired feature as it is directly associated with tool fragility or brittleness. High hardness increases the sustainability of the tool against permanent change in shape and geometry during machining; while, it consequently lowers the fracture strength or toughness during impacts.

Throughout its service life, a cutting tool is subjected to different types of loading, unloading, vibration and other interfering factors. A successful candidate surviving these situations is the one who absorbs the energy imposed by cyclic forces and vibrations without showing any signs of fracture. This capability is generally refer to as toughness which is the ability of a cutting tool to absorb energy before fracture. Figure 2.5 shows the hardness versus toughness for some commonly used cutting tool material.

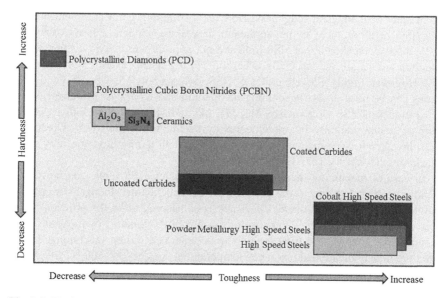

Fig. 2.5 Hardness versus toughness for some conventional cutting tool materials

Another desired characteristic that a successful cutting tool must possess is wear resistance. Wear is normally defined as the erosion of tool particles by means of another moving surface. Based on the definition of wear, wear resistance can be defined as the ability of cutting tool material to retain its integrity against erosion and eventually demonstration of acceptable tool life.

A desired cutting tool for particular application is the one who demonstrates a balanced combination of all aforementioned features. The question to be addressed here is what types of cutting tools are appropriate for machining titanium and its alloys.

2.6 Application of HSS Tools in Machining Titanium and Its Alloys

High Speed Steel (HSS) tools have been extensively employed for machining of a different kind of materials through the past decades. HSS tools show great toughness in comparison to other tool materials and capable of withstanding against the cyclic or intermittent loading and unloading. For this reason, they are primarily used for cutting operations at which interrupted or intermittent cutting is likely to occur. These include operations such as milling, drilling, broaching, and tapping. High speed steels are also appropriate for producing tools of complex shape such as helical milling, drilling, broaching, reaming, and tapping tools. However, when it comes to machining of hard-to-cut materials such as titanium or even tempered steels, HSS tools are not the best option to select.

In spite of their great toughness, HSS tools softening point is around 600 °C [35, 36]; therefore, there are not applicable in working temperatures above 500 °C [33]. It has been shown that HSS tools are not appropriate options for machining titanium and its alloys; especially, when the cutting speeds exceed 30 m/min [37, 38]. However, highly alloyed grades of HSS tools such as T5, Tl5, M33 and M40 series can be used toward machining of titanium and its alloys. Moreover, some other grades of HSS tools such as M1, M2, M7 and Ml0, which are called general purpose grades, can also be implemented in machining titanium. [27]. Great care must be taken to keep the cutting speed lower than 30 m/min limit specially when machining Ti-6Al-4V [27, 39].

In case of machining titanium and its alloys using HSS tools, cutting tool is rapidly deteriorated due to the presence of several factors among them, plastic deformation have the dominant effect and is considered to be the principal wear mechanism [7, 26]. Plastic deformation is principally originated by means of high compressive stresses and high temperatures generated during machining. It has been previously mentioned that when temperatures goes beyond 600 °C which is a common case when machining titanium alloys, thermal softening of HSS tools starts. Hence, HSS tools considerably lose their hardness which consequently accelerates plastic deformation at and around the tool nose or cutting edge. Plastic deformation of the tool nose and cutting edge makes the tool no longer functional. Another temperature related type of wear that has been observed when machining titanium alloys using HSS tools is the crater wear caused by intense temperature created while the chip is moving over the rake face [26, 40]. The combined effect of plastic deformation and crater wear rapidly destroy the HSS tools and thus makes it inappropriate choice for machining titanium and its alloys.

Although the performance of HSS tools can be further improved by application of coating, not all of those coated HSS tools can be used toward machining titanium and its alloys. Titanium nitride (TiN) and titanium carbonitride (TiCN) coated HSS tools are not suggested due to their chemical affinity with titanium. Their tendency to chemically react with titanium rapidly destroys the coating and leaves the tool surface unprotected. Other commonly used coatings for HSS tools such as chrome nitride (CrN), and titanium aluminum nitride (TiAlN) appear to be more advantageous in machining titanium alloys. However, HSS tools are not the material of choice for machining titanium and its alloys.

2.7 Application of Carbide Tools in Machining Titanium and Its Alloys

The term carbide tool is referred to a broad range of cutting tools that are made from carbide particles using different production methods [32, 41]. As previously mentioned in this chapter, carbide tools are mixtures of hard carbide particles (WC, TiC, TaCx, NbC, and Nb2C) bounded together in a matrix of cobalt [33]. These tools are also referred to in industry as sintered carbide, cemented carbide or

Table 2.2 Effects of additive materials on the characteristics of carbide tools

	Positive effects	Negative effects
Cobalt (W)	Increase toughness and shock resistance	Decreases hardness and wear resistance
Tungsten carbide (WC)	Increase hardness and wear resistance	Decrease toughness
Titanium carbide (TC)	Increases wear resistance	Decrease toughness
Tantalum carbide (TaC)	Increases hot hardness and prevent plastic deformation	

hard-metal as they made from the combination of hard carbide particles together. These cutting tools are either manufactured directly from a block of raw material by grinding or in the form of small inserts with specific geometry [33].

The commercially available carbide tools in the market are categorized either as straight grade carbides or as mixed grade carbides or mixed grade carbides. The former is composed of 6 wt% cobalt (Co) and 94 wt% tungsten carbide (WC) while the latter can be achieved by incorporating additive elements such as titanium carbide (TiC), tantalum carbide (TaC) or niobium carbide (NbC) [36]. In comparison to the tungsten carbide (WC), titanium carbide (TiC) demonstrates greater hardness reaching up to 3,200 HV [33] and it is mainly utilized to promote the wear characteristics of carbide tools [36]. Increasing the TiC content improves the wear resistance; but, it negatively affects the toughness and fracture strength of carbide tools. Although carbide tools have higher hot hardness than the carbon steel and high speed steel tools, their capability to endure high temperatures can be further improved by adding tantalum carbide. As a result, the machining can be performed at higher cutting speed without any concern about plastic deformation of the cutting edge due to resultant temperatures. The following table shows a brief summary of the effects of additive materials on the characteristics of carbide tools (Table 2.2).

Capability of carbide tools to resist against diffusion and oxidation wear along with their hot hardness can be improved by adding TiC, TiN, Al_2O_3, TiCN, TiAlN, TiZrN, TiB2 and diamond coatings.

One of the preferred carbide tools for machining titanium and its alloys is the straight grade cemented carbide (WC–Co) comprising 6 wt% cobalt (Co) and tungsten carbide (WC) grain size whiting the range of 0.8–1.4 μm [7, 27, 39, 42–44]. However, cutting speeds in excess of 60 m/min is not suggested for cemented tungsten carbide tools [44]. Application of higher cutting speeds (>60 m/min) is normally confined by plastic deformation due to intense heat [43]. Machining titanium and its alloys with low cutting speed (<45 m/min) eliminates the effects of thermal softening and also reduces the possibility of chemical reaction between tool and workpiece. As a result, mechanical and thermal fatigue, as well as micro-fractures is the dominant failure modes in such cases.

Increasing the cutting speed will result in elevating the temperature at the cutting zone. Due to the poor thermal conductivity of titanium, this intense heat remains at the cutting zone and elevates the temperature which may reach 500 °C and even higher [36]. The titanium becomes highly reactive and initiation of diffusion wear is likely to occur at such temperatures. In such cases, the titanium atoms migrate from the workpiece and chemically react with the carbon content of carbide tools. This chemical reaction produces titanium carbide (TiC) [36, 39] which is extremely hard and it rigorously adheres to both the workpiece and tool. This strongly adhered layer of titanium carbide protects the tool from further diffusion.

Another type of wear when machining titanium and its alloys using carbide tools is adhesion wear which is likely to occur in temperatures above 740 °C in the presence of normal contact pressure of 0.23 GPa [45]. The tool material is repeatedly damaged and consequent adhesion wear is accelerated when these welded particles are detached by the flow of the chip over the rake face.

The grain size has a great effect on the wear resistance of carbide tools. The carbide tools with coarser grains show higher crater wear rate due to the fact that coarser grains are more susceptible to be pulled off when the chip slides over the rake face. The resistance of carbide tools against the crater wear can be improved by reducing the grain size; however, the carbide tolls with smaller grain size demonstrates higher flank wear rate [46].

The coated carbide tools are generally produced using either CVD (chemical vapor deposition) or PVD (physical vapor deposition) techniques. CVD coatings greatly adhered to the carbide tools and improve the wear resistance. PVD coating also increase the wear resistance as well as edge toughness.

Application of coating reduces friction at tool-chip interface; hence, it lowers the cutting forces and also heat generated due to friction. In addition, the coating layer act as a shield and it protects the tool from thermal shocks during machining. However, the rapid increase in wear rate has been observed when the coating layer is removed after several pass of cutting and the tool is directly exposed to the workpiece. The situation worsens when machining titanium and its alloys using coated carbides. In this case, the coating layer is rapidly removed either by abrasion wear due to the fast flow of chip or by the chemical reaction between coating and titanium workpiece. Coating materials such as titanium nitride (TiN) or titanium carbonitiride (TiCN) are vulnerable to chemical reaction with titanium and their application in machining titanium must be prohibited [47]. Research studies have proven that straight-grade cemented carbide tools with no coating exhibit better performance than those coated by TiC, TiN–TiC, AI_2O_3–TiC, TiN–Ti(C,N)–TiC, AI_2O_3 gamma layer, HfM, and TiB_2 [26]; hence, straight grade cemented carbide (WC–Co) with no coating is one of the most preferred carbide tools in machining titanium and its alloys.

2.8 Application of Ceramic Tools in Machining Titanium and Its Alloys

Ceramic tool demonstrates several promising and unpromising features simultaneously. They demonstrate much higher compressive strength than their HSS and carbide counterparts; however, they are very vulnerable to mechanical and thermal

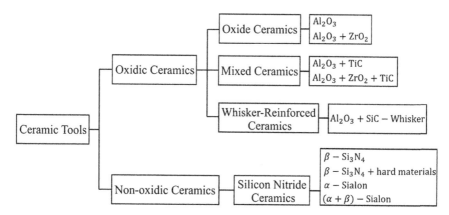

Fig. 2.6 Classification of ceramic tools

shock during machining and they easily break under heavy or interrupted machining due to brittleness and lack of toughness. In addition to great compressive strength, ceramic tools offer high hot hardness, chemical inertness and better oxidation resistance [48]. Ceramic tools owe their hot hardness to their high melting point and absence of binder as the second phase.

Ceramic as a cutting tool material can be divided into two main classifications and four sub categories. These categories are presented in the Fig. 2.6.

As can be seen, two main categories of ceramics tools are oxide and non-oxide ceramics. The oxide ceramic family which includes pure oxide, mixed oxide and whisker-reinforced ceramics is also known as alumina based family as the base material in all of them is aluminium oxide or alumina (Al_2O_3). The non-oxide ceramic family mostly includes silicon nitride-based ceramics.

The pure oxide ceramic tools based on aluminium oxide (Al_2O_3) are very brittle and vulnerable to fracture; hence, their fracture toughness is normally improved by adding zirconium oxide (ZrO_2) [49]. Zirconium oxide increases the fracture toughness with no negative effect on wear resistance. However, this category of ceramic tools is barely used in machining difficult to cut materials due to their low thermal shock resistance and low fracture toughness at elevated temperatures.

In order to further improve the characteristics of ceramic tools, pure ceramic can be mixed with 5–40 % [49] of titanium carbide (TiC) or titanium carbonitride (TiCN). The obtained ceramic is called mixed oxide ceramic, which exhibits comparatively higher hot hardness, higher hardness and thermal shock resistance. However, adding titanium carbide adversely affects the fracture toughness of the tool.

The performance of ceramic tools can be even further enhanced by implementing 20–40 wt% silicon carbide whisker into the Alumina (Al_2O_3) as a base material [49]. This family of ceramic tools is known as whisker-reinforced ceramic tools. Silicon carbide whiskers act as reinforcement and remarkably increase the toughness of ceramic tools. Whisker-reinforced ceramic tools demonstrate 60 % higher fracture toughness than mixed oxide ceramics tools [49].

Despite their promising characteristics, lack of toughness and consequently low fracture strength make ceramic tools inappropriate choice for machining titanium and its alloys. Ceramic tools show a higher rate of rake and flank face wear in comparison to the straight grade cemented carbides [43]. However, among ceramic tools, sialons demonstrate relatively higher resistant to rake face wear, in comparison to the Al_2O_3-TiC and Al_2O_3-30ZrO_2 [43].

Generally speaking, due to their fragility and lack of toughness, ceramic tools greatly suffer from notch wear and large groove wear when utilized toward machining of titanium and its alloys [44]. The notch wear is primarily initiated by the fracture process caused by cyclic forces generated during the formation of discontinues serrated chips. For these reasons, similar to HSS tools, ceramic are also not an appropriate option for machining titanium and its alloys.

2.9 Application of CBN Tools in Machining Titanium and Its Alloys

Boron nitride exists in the nature in the form of hexagonal boron nitride which is a soft material. In its hexagonal modification, boron nitride does not present the required characteristics to be considered as an appropriate cutting tool material. In order to achieve the desired characteristics, the hexagonal modification must be transferred into the cubic crystalline lattice. This process takes place by applying high-pressure and also high temperature. After the transformation of hexagonal boron nitride into cubic boron nitride, it reveals its superior characteristics as a cutting tool material.

CBN has the second highest hardness among the tool materials (after diamond) and also has a high melting point (2,730 °C); as a result, it shows outstanding high hot hardness in elevated temperatures. In addition, despite diamond that begins to graphitize already at about 900 °C, CBN shows superior oxidation resistance and it remains stable up to 2,000 °C with no sign of oxidation [49].

The crystal size of cubic boron nitride (CBN) is very small (1–50 μm); therefore, CBN grains are sintered together by means of binder under high temperature and pressure to form polycrystalline cubic boron nitride. This family of CBN tools is called PCBN tools [49].

Widely respected due their unique characteristics, CBN and PCBN tools are primarily used in the machining of hard-to-cut materials. Machining of forged steels (45–68 HRC), alloy steels (70 HRC), and nickel and cobalt based super alloys [50]. They are also the dominant tools for hard turning where high metal removal rate and acceptable surface roughness must be achieved simultaneously. Nose deformation of CBN tools has been reported when machining $(\alpha + \beta)$ titanium alloys with 4.5 % aluminum (Al) and 4.5 % manganese (Mn) [51]. It has also been reported [51] that the combined effect of elevated temperature and compressive stress may reached a point at which the tool is no longer able to sustain; thus, deformation and the wear initiate. Low cycle fatigue caused by the cyclic mechanism of chip formation, nose wear, chipping, and diffusion wear in the presence of nitrogen and oxygen are among other types of wear when CBN is used toward machining titanium and its alloys.

In conclusion, although CBN and PCBN tools are capable of reaching higher cutting speed (150 m/min for Ti–6Al–4V [52] and 185–220 m/min for $\alpha + \beta$ [51]) than carbide tools; they are not very popular in machining titanium and its alloys. This is mainly due to their expensive price that could be 10–20 times higher than their carbide counterparts and make their application not economically efficient.

2.10 Application of PCD Tools in Machining Titanium and Its Alloys

Diamond is the hardest existing material on the earth and is much harder than silicon carbide (SiC) and tungsten carbide (WC). In addition to extreme hardness, diamond also shows good resistance to wear and low coefficient of friction which give it numerous advantages over other types of cutting tools especially the abrasive ones. As manmade tools, PCDs are the combination of diamond particles (crystals) of different size bonded together by means of metallic bonder (usually cobalt). Depending on their grain size, PCD tools can demonstrate different characteristics and consequently different applications.

PCD tools with larger diamond particles (coarser grains) shows remarkable resistance to wear. However, coarser grains result in a tool with a rougher cutting edge, and consequently lead to lower a surface finish and a higher roughness of the workpiece surface. In contrast, the smaller diamond particles (finer grains) result in sharper edge and eventually higher surface finish but reduced tool life due to wear [32, 33]. To strike a balance between tool life and surface quality, general purpose PCD tools are usually made from medium size diamond particles to achieve an acceptable level of wear resistant as well as surface quality.

It has been observed that when applied for machining titanium alloys, PCD tools show much lower wear rate than carbide tools [53]. They also show acceptable performance in machining Ti-6Al-4V [54] which is among the most extensively utilized alpha-beta ($\alpha + \beta$) titanium alloys for producing compressor blades in aerospace industries. The lower wear rate of PCD tools in machining titanium alloys can be attributed to the formation of titanium carbide as a protective layer on the rake face of PCD tools due to the inter-diffusion of titanium and carbon. The hard layer of titanium carbide strongly adhered to the substrate and act as a barrier and prevents further diffusion of the tool material into the chip [47].

In general, if selected for proper applications, PCD tools have an acceptable performance in machining titanium alloys. However, the high tooling cost imposes a great barrier to their widespread application in this field.

2.11 Conclusion

Comprising high strength to weight ratio, corrosion resistance, fatigue resistance, and capability to work in high temperature, titanium is usually the primary candidate for aerospace, power generation, automotive and even medical industries. However, the widespread application of titanium is limited due to several reasons among them; the price and machinability are outstanding. Titanium has low modulus of elasticity, poor thermal conductivity, chemical reactivity, and hardening characteristics that together make it one of the most notorious materials to machine. Due to the above-mentioned characteristics, an appropriate cutting tool for machining titanium and its alloys should possess several characteristics to be considered as an ideal candidate. These characteristics include but not limited to the capability of maintaining hardness at high cutting temperatures, high toughness, resistance to cyclic loading and unloading, and also chipping resistance. It should also show low chemical affinity with titanium and also good thermal conductivity to dissipate the heat generate during cutting. Different cutting tool materials are candidates for machining titanium alloys; however, each of them exhibits some signs of limitation. Dramatic loss of hardness at high temperature makes HSS tools not good candidates for machining titanium and its alloys. Carbide tools are among the most widely used cutting tools in machining titanium and its alloys due to their comparatively acceptable combination of hardness and toughness. Although ceramic tools have high hardness and low chemical affinity with titanium, they are not appropriate for machining titanium and its alloys, due brittleness and lack of toughness. CBN tools are very vulnerable to fracture and chipping primarily because of their extreme hardness. Consequently, their application as a cutting tool in machining titanium is confined to finishing operations. PCD tools are among other appropriate but expensive tools for machining titanium and its alloys. Although carbon content of the PCD tools is likely to react with titanium, this process is being eliminated by the formation of titanium carbide layer which strongly sticks to the tool and protect it from further diffusion wear.

References

1. Britannica E (2001) Encyclopaedia Britannica Online. Encyclopædia Britannica
2. Gerdemann SJ (2001) Titanium process technologies. Adv Mater Process 159(DOE/ARC-2002-007)
3. Jackson M, Dring K (2006) A review of advances in processing and metallurgy of titanium alloys. Mater Sci Technol 22(8):881–887
4. Titanium (2014) http://www.chemguide.co.uk/inorganic/extraction/titanium.html
5. McQuillan AD, McQuillan MK (1956) Titanium, vol 4. Academic Press, New York
6. Habashi F (1997) Handbook of extractive metallurgy, vol 2. Wiley-Vch, Weinheim
7. Ezugwu E, Wang Z (1997) Titanium alloys and their machinability—a review. J Mater Process Technol 68(3):262–274
8. Joshi VA (2006) Titanium alloys: an atlas of structures and fracture features.CRC Press, Boca Raton

9. Yang X, Liu CR (1999) Machining titanium and its alloys. Mach Sci Technol 3(1):107–139
10. Voskoboinikov R, Lumpkin G, Middleburgh S (2013) Preferential formation of Al self-interstitial defects in γ-TiAl under irradiation. Intermetallics 32:230–232
11. Groover MP (2007) Fundamentals of modern manufacturing: materials processes, and systems.John Wiley & Sons, Hoboken
12. Astakhov VP, Xiao X (2008) A methodology for practical cutting force evaluation based on the energy spent in the cutting system. Mach Sci Technol 12(3):325–347
13. Astakhov VP (2006) Tribology of metal cutting, vol 52.Accessed Online via Elsevier
14. Outeiro J (2003) Application of recent metal cutting approaches to the study of the machining residual stresses. Department of Mechanical Engineering, University of Coimbra, Coimbra p 340
15. Child H, Dalton A (1968) ISI special report 94. London
16. Konig W (1978) Applied research on the machinability of titanium and its alloys. In: Proceedings of AGARD conference advanced fabrication processes, Florence
17. Fitzsimmons M, Sarin VK (2001) Development of CVD WC–co coatings. Surf Coat Technol 137(2):158–163
18. Pramanik A (2014) Problems and solutions in machining of titanium alloys. Int J Adv Manuf Technol 70(5–8):919–928
19. He G, Zhang Y (1985) Experimental investigations of the surface integrity of broached titanium alloy. CIRP Ann-Manuf Technol 34(1):491–494
20. Donachie Jr M (1982) Introduction to titanium and titanium alloys. American Society for Metals, Titanium and Titanium Alloys Source Book, p 7
21. Barry J, Byrne G, Lennon D (2001) Observations on chip formation and acoustic emission in machining Ti–6Al–4 V alloy. Int J Mach Tools Manuf 41(7):1055–1070
22. Komanduri R, Von Turkovich B (1981) New observations on the mechanism of chip formation when machining titanium alloys. Wear 69(2):179–188
23. Vyas A, Shaw M (1999) Mechanics of saw-tooth chip formation in metal cutting. J Manuf Sci Eng 121(2):163–172
24. Sun S, Brandt M, Dargusch M (2009) Characteristics of cutting forces and chip formation in machining of titanium alloys. Int J Mach Tools Manuf 49(7):561–568
25. Komanduri R, Hou Z-B (2002) On thermoplastic shear instability in the machining of a titanium alloy (Ti-6Al-4V). Metall Mater Trans A 33(9):2995–3010
26. Machado A, Wallbank J (1990) Machining of titanium and its alloys—a review. Proc Inst Mech Eng B J Eng Manuf 204(1):53–60
27. Siekmann HJ (1955) How to machine titanium. Tool Eng 34(1):78–82
28. Stoughton B (1908) The metallurgy of iron and steel. McGraw-Hill publishing company, New York
29. Oberg E, Jones FD (1918) Iron and steel: a treatise on the smelting, refining, and mechanical processes of the iron and steel industry, including the chemical and physical characteristics of wrought iron, carbon, high-speed and alloy steels, cast iron, and steel castings, and the application of these materials in the machine and tool construction. The Industrial Press, New York
30. Society of Manufacturing Engineers (2014) https://manufacturing.stanford.edu/processes/CuttingToolMaterials.pdf
31. Machinist (2014) http://americanmachinist.com/cutting-tools/chapter-1-cutting-tool-materials
32. Davim JP (2008) Machining: fundamentals and recent advances. Springer, London
33. Davim JP (2011) Machining of hard materials. Springer, London
34. Whitney ED (1994) Ceramic cutting tools. William Andrew Pub, Norwich
35. Kramer B (1987) On tool materials for high speed machining. J Eng Ind 109(2):87–91
36. Ezugwu E, Bonney J, Yamane Y (2003) An overview of the machinability of aeroengine alloys. J Mater Process Technol 134(2):233–253
37. Rahman M, Wang Z-G, Wong Y-S (2006) A review on high-speed machining of titanium alloys. JSME Int J, Series C 49(1):11–20

38. Pérez J, Llorente J, Sánchez J (2000) Advanced cutting conditions for the milling of aeronautical alloys. J Mater Process Technol 100(1):1–11
39. Hartung PD, Kramer B, Von Turkovich B (1982) Tool wear in titanium machining. CIRP Annal Manuf Technol 31(1):75–80
40. Trent EM, Wright PK (2000) Metal cutting. Butterworth-Heinemann, Boston
41. Upadhyaya G (1998) Cemented tungsten carbides: production, properties and testing. William Andrew
42. Che-Haron C (2001) Tool life and surface integrity in turning titanium alloy. J Mater Process Technol 118(1):231–237
43. Dearnley P, Grearson A (1986) Evaluation of principal wear mechanisms of cemented carbides and ceramics used for machining titanium alloy IMI 318. Mater Sci Technol 2(1):47–58
44. Narutaki N et al (1983) Study on machining of titanium alloys. CIRP Annal Manuf Technol 32(1):65–69
45. Sadik MI, Lindström B (1995) The effect of restricted contact length on tool performance. J Mater Process Technol 48(1):275–282
46. Edwards R, Edwards R (1993) Cutting tools. Institute of Materials, London
47. Rahman M, Wong YS, Zareena AR (2003) Machinability of titanium alloys. JSME Int J Series C 46:107–115
48. Chakraborty A, Ray K, Bhaduri S (2000) Comparative wear behavior of ceramic and carbide tools during high speed machining of steel. Mater Manuf Process 15(2):269–300
49. Klocke F (2011) Manufacturing processes. Springer, Berlin
50. Lin Z-C, Chen D-Y (1995) A study of cutting with a CBN tool. J Mater Process Technol 49(1):149–164
51. Zoya Z, Krishnamurthy R (2000) The performance of CBN tools in the machining of titanium alloys. J Mater Process Technol 100(1):80–86
52. Ezugwu E et al (2005) Evaluation of the performance of CBN tools when turning Ti–6Al–4V alloy with high pressure coolant supplies. Int J Mach Tools Manuf 45(9):1009–1014
53. Nabhani F (2001) Machining of aerospace titanium alloys. Robot Comput Integr Manuf 17(1):99–106
54. Kuljanic E et al (1998) Milling titanium compressor blades with PCD cutter. CIRP Annal Manuf Technol 47(1):61–64

Chapter 3
Mechanics of Titanium Machining

Ismail Lazoglu, S. Ehsan Layegh Khavidaki and Ali Mamedov

Abstract Titanium is widely used material in advanced industrial applications such as in aeronautics and power generation systems because of the distinguished properties such as high strength and corrosion resistance at elevated temperatures. On the other hand, the machinability of this material is poor. Relatively low thermal conductivity of Titanium contributes to rapid tool wear, and as a result, high amounts of consumable costs occur in production. Therefore, understanding the mechanics of titanium machining via mathematical modeling and using the models in process optimization are very important when machining Titanium both in macro and micro scales. In this chapter, mechanical effect of process parameters in five axis milling and micro milling are analyzed. Thus, different cutting conditions were tested in dry conditions and the effects of tool orientation on cutting forces in five axis macro milling was investigated. For five-axis ball end milling operation, a series of experiments with constant removal rate and different tool orientation (different lead and tilt angle) were conducted to investigate the effect of tool orientation on cutting forces. The aim of the tests was finding the optimum orientation of the cutter in which the normal cutting force applying on machined surface is minimum. Moreover, a new method to predict cutting forces for micro ball end mill is presented. The model is validated through sets of experiments for different engagement angles. The experiment and the simulation indicated that the tool orientation has a critical effect on the resultant cutting force and the component that is normal to the machined surface. It also possible to predict the tool orientation in which the cutting torque and dissipated energy is minimum. In micro milling case, the force model for ball end mill is able to estimate the cutting forces for different cutting conditions with an acceptable accuracy.

I. Lazoglu (✉) · S. Ehsan Layegh Khavidaki · A. Mamedov
Manufacturing and Automation Research Center, Koc University,
Sariyer, 34450 Istanbul, Turkey
e-mail: ilazoglu@ku.edu.tr

J. P. Davim (ed.), *Machining of Titanium Alloys,*
Materials Forming, Machining and Tribology, DOI: 10.1007/978-3-662-43902-9_3,
© Springer-Verlag Berlin Heidelberg 2014

3.1 Introduction

Ti-6A1-4V is an advanced engineering alloy finding wide applications. High
strength, corrosion and fatigue resistance at elevated temperatures are superior
properties of this alloy. Ti-6Al-4V is used in various industries such as aeronautics,
nuclear energy generation plants, food processing machinery, biomedical product
and implant manufacturing. The reason that this alloy is widely used in aeronautics
is the superior characteristics that make it suitable to be used in extreme working
conditions such as jet engines. Jet engines capable of running at high temperatures
work more effectively and fuel consumption decreases; in addition components
designed with this alloy is providing air vehicles advantage in fuel consumption due
to being smaller and lighter than the components designed with traditional steels
[1]. Although having superior properties, Ti-6Al-4V has poor machinability.

Low thermal conductivity of Ti-6Al-4V causes high cutting temperature. High
strain hardening ability and high reactivity with cutting tool materials of this alloy
result in rapid tool wear at elevated temperatures. Ti-6Al-4V has low modulus of
elasticity and high tendency in decrease of modulus of elasticity with increasing
temperature which causes deflection during thin wall machining [2]. Another
problem in machining of titanium alloys is serrated chip formation even in low
cutting speeds. The main reason for this behavior is the exhibition of instability
during plastic deformation; locally increased temperature in shear zone decreases
the flow stress and as a result, deformation occurs on a thin plane. This periodic
serrated chip formation yields in fluctuation in the cutting forces and excitation of
periodic cutting forces result in chatter phenomenon [3].

The problems occur during the machining of this alloy increase manufacturing
costs, decrease product quality and efficiency. Within this perspective, studies
aiming to optimize machining process are becoming more important. Many
researchers have studied about machining of Ti-6A1-4V. Generally, these studies
are based on mechanistic analysis. Since having tool wear problem is an important
characteristic of these alloys in machining, wear mechanism is investigated in
most of these studies. Among these researches, presenting only wear examination
[2, 4, 5], relation between wear and cutting forces examination [6−8], comparison
of different cutting tools or coating performance examination [9, 10], chip formation, wear and cutting coefficients examination [11], only cutting forces
examination [12], different tool geometry effects on the surface properties and
cutting forces examination [13] studies are available. In some studies temperature
analysis is also performed. Among these researches, presenting analysis of temperature and tool wear and determining the correlation by executing experimental
and modeling studies simultaneously [14] analysis of temperature, wear, cutting
forces and the type of chip formation together [15] thermo mechanical analysis
based on experimental and finite elements method [16], force, temperature simulations and examination of chip types [17] studies have been published. In
general, there is a lack of study considering both thermal and mechanical effects in
machining of Ti-6A1-4V.

Ball end milling of Titanium alloy have not investigated widely. The ball end milling operation is used extensively in manufacturing of free form surfaces in hitech industries such as aerospace, die/mold, automotive and bio medical implants. Due to the different cutting speed along the tool axis, ball end milling operation is extremely depended upon the cutting parameters such as feed rate, tool orientation, spindle speed, etc.

Ng et al. [18] conducted some experiments to investigate the effects of cutter orientation, tool coating and cutting environment on tool life, tool wear mechanisms, cutting forces, chip formation, cutting temperature and workpiece surface roughness. In this study, rough cutting is considered during horizontal downward orientation of cuter with using high pressure cutting fluid. According to this research the tool life can be increased up to 15 m cut length which is two times longer than dry cutting. Also, it is concluded that machining at a workpiece angle of 45°, horizontal downwards cutter orientation generated the longest length cut.

Surface integrity in ball end milling of Titanium alloys has been studied by Mahamdi et al. [19] and Joshi et al. [20]. Based on those researches the root mean square roughness (sq) increases from 2.9 micron to 4.8 micron by increasing the feed speed from 300 mm/min to 900 mm/min for ball end milling of Titanium alloy. It is also declared that the tool positioning can affect the sq up to 50 %.

Despite all of the above mentioned studies, the effects of lead and tilt angle in ball end milling of super alloys has not investigated comprehensively. Due to the complexity of calculation of engagement between the cutter and workpiece in machining of free form surfaces, simulation of the process and predicting the optimum tool orientation is still unknown.

Micro milling is a commonly used micro machining method in manufacturing of miniature biomedical parts, micro sensors and actuators, micro dies and molds. Micro milling is widely used process due to several advantages like flexibility of the process, which allows manufacturing freeform 3D complex geometries and high material removal rate, compared to other micro manufacturing techniques. Micro milling process has its difficulties as well. The main difficulty is fragile micro tools. This aspect is becoming more essential when workpiece material is hard to machine like titanium alloys. Due to the low thermal conductivity of these alloys higher portion of the temperature generated during cutting process flows into the cutting tool which causes excessive wear and premature failure of micro tools. Several groups of researchers are working on micro milling of Titanium to understand process kinematics, improve machinability and decrease tool wear during cutting process. The most effective method to decrease tool wear is coating of micro tools. The prior research on wear mechanisms of polycrystalline diamond (PCD), cubic crystalline boron nitride (cBN) and TiB_2 coated carbide tools during high performance machining of Ti-6Al-4V titanium alloy is done by Corduan et al. [21]. Later Aramcharoen et al. [22] compared different coatings like TiN, TiCN, TiAlN, CrN and CrTiAlN and their effect on flank wear, chipping, edge radius wear, burr size and machined surface quality. Ozel et al. [23] investigated effects of CBN, TiAlN and TiAlN + cBN coatings on wear rate distribution and cutting forces formed during micro milling process. Thepsonthi et al. [24] investigated the effect of cBN coating on cutting temperature and stated that the tool temperature and wear rate increase by increasing the cutting speed and feed rate.

The critical factor in micro milling is to manufacture high precision parts within a tight surface and form tolerances. It is known that portion of dimensional, surface and form errors are induced by deflections of micro tool caused by cutting forces. Due to this reason, many researchers worked on modeling of kinematics of the micro milling cutting process. In foregoing research Vogler et al. [25] and Waldorf et al. [26] have developed shear plane plasticity model. Based on these studies June [27] and Fang et al. [28] have developed more complex plasticity models, which covers elastic recovery of plowed material. Fang [29] proposed a generalized slip-line field model to predict shearing and plowing forces. Rodriguez and Labarga [30] developed an analytical force model for micro end milling considering the size effect and eccentric deviation of the tool path. Jin and Altintas [31] presented a slip-line field model, which considers the stress variation in shear zones and relates it to temperature and tool edge radius effect. A mechanistic force model proposed by Park and Malekian [32] considers both the shearing and plowing dominant cutting regimes and relates plowed area to volume of the plowed material due to the effect of tool edge radius. But all presented above work is done for flat end mill.

In this research, mechanical effects of machining parameters in ball end milling and a new mathematical model to predict cutting forces during micro milling with ball end mill processes are presented. Estimated cutting forces are presented as a function of cutting coefficients. A comprehensive chip thickness model developed by Li et al. [33] for micro milling is used for precise estimation of cutting forces. The predicted micro milling forces are validated by the experimental tests and some of the results are shared in this chapter. A cobalt coromill plura from Sandvik was used for macro and micro ball end milling of Ti-6A1-4V.

3.2 Mechanics of Orthogonal Cutting

First step to investigate kinematics of ball end milling cutting process is to estimate cutting forces. There are two main ways to predict the cutting forces: mechanistic calibration and orthogonal to oblique transformation. The advantage of mechanistic calibration method is its accuracy; however, this method is not general and requires various cutting tests with different chip load and feed rate. On the other hand, orthogonal to oblique transformation method can be implemented to any tool geometry and is not dependent on cutting conditions. For five axis macro ball end milling operation orthogonal to oblique transformation method was employed, but for micro ball end milling in order to have more accurate force prediction mechanistic calibration method was used.

Orthogonal cutting process can be used in the mechanistic modeling of chip removal processes for machining of all metals and alloys. This approach is more convenient, since it has very simple geometry comparing to 3D oblique process.

Fig. 3.1 Mechanics of orthogonal cutting

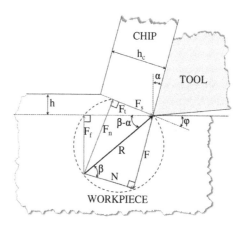

The cutting tool should have a flat surface and a straight edge in the orthogonal cutting process. For practical application of orthogonal cutting process, it is assumed that the cutting edge has no chamfer or radius. Besides, the chip removal process is assumed to have ideal conditions along the whole cutting edge when the tool has been positioned perpendicular to the cutting velocity. Therefore, this process can be defined as deformation without side spreading [34]. Since chip formation occurs in two dimensions, the cutting forces occur just in feed and cutting speed directions. Mechanics of orthogonal cutting is shown in Fig. 3.1.

Cutting process occurs with the penetration of cutting tool into workpiece, forming of cutting region in front of the cutting edge, and flowing of workpiece material along the rake face. In this study, the cutting zone has been assumed to be infinitely thin plane. The shear plane has been located on the adjacent planes of tool cutting edge and workpiece material surface. The shear angle, φ, is defined as the angle between shear plane and the cutting velocity. It is assumed that friction on the chip surface is constant during the cutting process such as the shear stress (τ_s) occurred in the shear plane, normal stress (σ_s) and forces acting on the shear plane have been balanced by the resultant force acting on the chip surface [35]. The effective force in the chip removal process has been separated in two components as the shear force (F_s) responsible from the chip removal and the normal force (F_n) applying a compressive stress on to shear plane. The following expressions are formulated when the resultant force (R) on the contact point of tool-workpiece material is related with the geometry shown in Fig. 3.1:

$$F_s = R \cos (\varphi + \beta - \alpha) \qquad (3.1)$$

$$F_n = R \sin (\varphi + \beta - \alpha) \qquad (3.2)$$

In these expressions, β represents the average friction angle and α represents the rake angle. In orthogonal cutting experiments, cutting forces are measured in two directions such as feed (F_f) and cutting speed (F_t) directions. The following geometrical relation is used to associate shear force with the above data:

Fig. 3.2 Orthogonal cutting-velocity relations

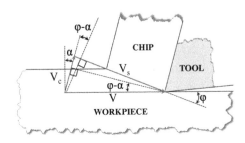

$$F_s = F_t \cos(\varphi) - F_f \sin(\varphi) \qquad (3.3)$$

The shear velocity can also be calculated geometrically in orthogonal cutting process. The relation between the cutting speed and shear velocity is illustrated in Fig. 3.2. This geometry is based on the fact that the vector sum of the relative velocity (V_c) of chip to the tool and the relative velocity (V) of the tool to the workpiece equals to relative velocity (V_s) of chip to the workpiece [35]. The following expression is obtained by using above relation:

$$V_s = V \frac{\cos(\alpha)}{\cos(\varphi - \alpha)} \qquad (3.4)$$

To calculate shear angle in the above expression by using the cutting geometry, the chip thickness need to be measured after performing orthogonal experiments. The shear angle is calculated by measuring deformed chip thickness and substituting that value in the below expression:

$$\varphi = \tan^{-1} \frac{h \cos(\alpha)}{h_c - h \sin(\alpha)} \qquad (3.5)$$

In this expression, h represents the depth of cut. After obtaining shear force and velocity, energy, occurred during the cutting process and reverts to heat, is calculated as follows to use in thermal analyze:

$$P_s = F_s \cdot V_s \qquad (3.6)$$

There is a secondary zone where the cutting forces are in equilibrium and energy dissipated due to the friction between chip and the tool. On the secondary zone, located on the tool chip surface, the resultant force acting on the cutting edge is balanced by the friction force on the rake face and normal force. These forces are expressed as follows (Fig. 3.2):

$$N = F_t \cos(\alpha) - F_f \sin(\alpha) \qquad (3.7)$$

Fig. 3.3 Orthogonal cutting
test set-up

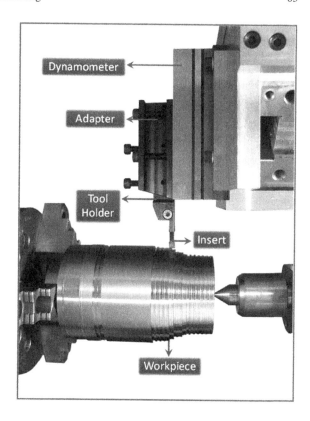

$$F = F_t \sin(\alpha) + F_f \cos(\alpha) \qquad (3.8)$$

The ratio of friction force to the normal force (F/N) gives the friction coefficient (μ) that is assumed to be constant through the process. By using the tool-workpiece geometry (Fig. 3.2), the velocity of the chip on the rake face can be calculated as follows:

$$V_c = V \frac{\sin(\varphi)}{\cos(\varphi - \alpha)} \qquad (3.9)$$

The energy occurred due to the friction on the chip and tool contact surface can also be calculated with similar approach to spend energy during cutting process:

$$P_s = F_s \cdot V_s \qquad (3.10)$$

There is a high effect of the energy, caused from friction, on the heat distribution of workpiece and tool.

Table 3.1 Orthogonal
database for Ti-6Al-4V [34]

$\tau_s = 613 \, (\text{Mpa})$
$\beta_a = 19.1 + 0.29\alpha_r \, (°)$
$r_c = C_0 h^{C_1}$
$C_0 = 1.755 - 0.028\alpha_r$
$C_1 = 0.331 - 0.0082\alpha_r$
$K_{te} = 24 \, (\text{N/mm})$
$K_{fe} = 43 \, (\text{N/mm})$

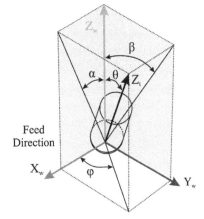

Fig. 3.4 Definition of lead
(α) and tilt (β) angles in five
axis milling

Test setup used in orthogonal cutting is illustrated in Fig. 3.3. Cutting forces are measured by a dynamometer consists of three-component force sensors.

Table 3.1 presents database obtained from mechanical analysis of orthogonal turning process of Ti-6Al-4V.

3.3 Five Axis Ball End Milling Operation

3.3.1 Kinematics of the Process

Despite of 3-axis milling, in five-axis milling, tool axis can have an inclination angle with respect to the normal vector to the XY plane of the machine table. As it is shown in Fig. 3.4, this inclination angle can be represented by tilt and lead angles. In this figure, $X_w Y_w Z_w$ is the workpiece coordinate frame and Z_t is the tool axis. Considering the feed direction, α and β are lead and tilt angles in Figs. 3.4, 3.5 respectively.

Since the local radius is varying along the tool axis in spherical part of the tool, cutting speed is not constant during the machining. At the tip of the tool the cutting

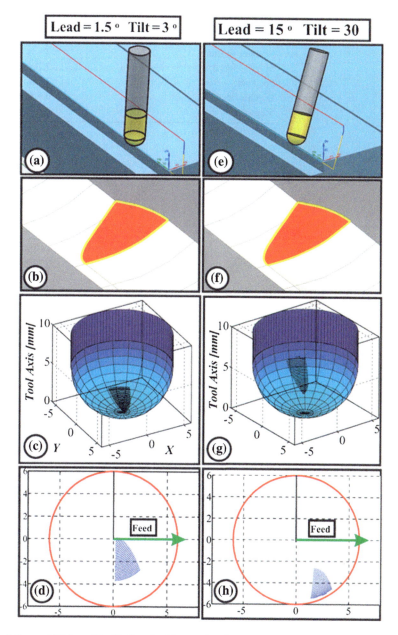

Fig. 3.5 **a**, **e** CAM model of ball end mill operation for two samples with different lead and tilt angle. **b**, **f** Map of engagement on workpiece. **c**, **g** Map of engagement on tool. **d**, **h** 2D map of engagement in order to calculate the entrance and exit angles

Table 3.2 Cutting conditions	Tool diameter	6 (mm)
	Over hang	30 (mm)
	Number of flutes	2
	Spindle speed	7,700 rpm
	Feed rate	1540 mm/min
	Depth of cut	0.5 (mm)
	Step over	0.5 (mm)

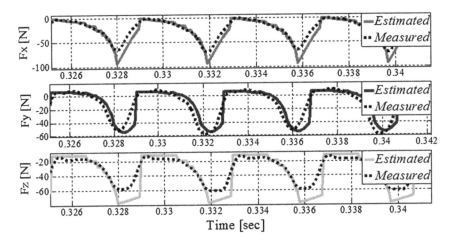

Fig. 3.6 Cutting forces for down milling Lead = 0 (deg.) and tilt = 0 (deg.)

speed is zero and the dominant phenomena are ploughing and indentation. However, by increasing the local radius, cutting speed increases.

Another difficulty in simulation of cutting process in five axis ball end milling is predicting the engagement between the tool and workpiece. In order to model the engagement map, an algorithm is developed using a solid modeler kernel.

3.3.2 Simulation and Experiment

Table 3.2 presents the cutting conditions and tool specifications in the simulation and the experiment.

The Cutting forces are simulated according to a mechanistic force model which is explained in [36]. In order to calculate the cutting force constants, the orthogonal to oblique method is employed by using the orthogonal database for Ti-6Al-4V which is mentioned in Table 3.1 [34]. In the table, τ_s is shear stress, β_a is friction angle, α_r is rake angle, h is uncut chip thickness, r_c is chip thickness ratio, K_{te} and K_{fe} are edge-cutting coefficients, respectively.

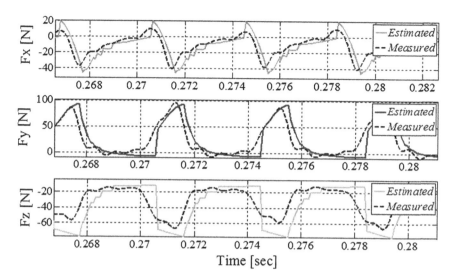

Fig. 3.7 Cutting forces for up milling Lead = 0 (deg.) and tilt = 0 (deg.)

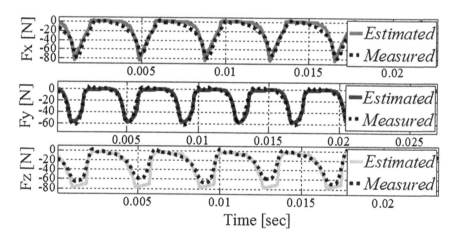

Fig. 3.8 Cutting forces for down milling Lead = 1.5 (deg.) and tilt = 3.7 (deg.)

Figures 3.6, 3.7, 3.8 and 3.9 illustrate the simulated and measured cutting forces. In all of the cases a very good agreement can be observed between the simulated cutting forces and measured forces. However, in z direction the maximum estimation error of 15 % is observed. The source of this error is mainly the ploughing and size effect phenomena during the ball end milling. This error can be reduces by using mechanistic calibration method instead of using orthogonal to oblique transformation method [37].

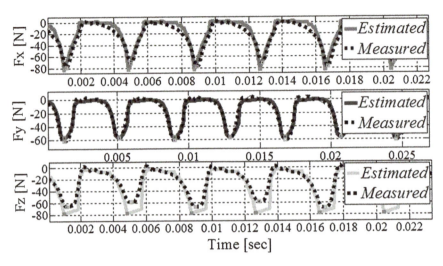

Fig. 3.9 Cutting forces for down milling Lead = 0 (deg.) and tilt = −4 (deg.)

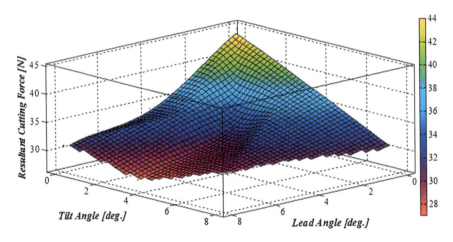

Fig. 3.10 Average resultant cutting force in down milling operation

3.3.3 Results

In order to investigate the effect of lead and tilt angle, series of tests were conducted with different lead and tilt angles. The average of the resultant cutting force and the normal component of the cutting force with respect to machined surface are considered and illustrated as 3D plots for down and up milling operations.

Figures 3.10 and 3.11 represent the average of resultant cutting force for different lead and tilt angles. It can be inferred from those figures that the maximum

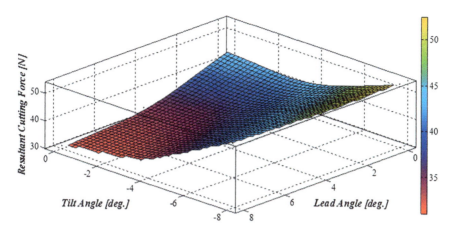

Fig. 3.11 Average resultant cutting force in up milling operation

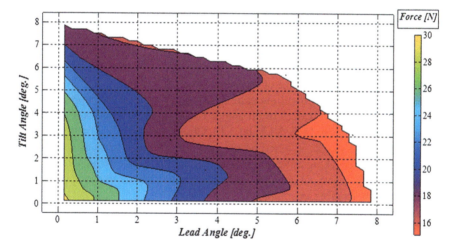

Fig. 3.12 Average of cutting force normal to machined surface for down milling operation

average resultant force occurs at zero lead and tilt angle for down milling. In contrast, the maximum resultant cutting force for up milling happens in zero lead angle and higher negative tilt values.

Form Fig. 3.10 it can be concluded that in order to have lower amount of resultant cutting force in down milling operation, the optimum tilt angle is around 4° and the lead angle should be selected more than 8°. For the up milling case, however, the optimum tilt angle is zero and the lead angle should be selected higher than 8°.

The surface integrity of finished workpiece is highly dependent on the component of machining force which is normal to final surface. Since the normal

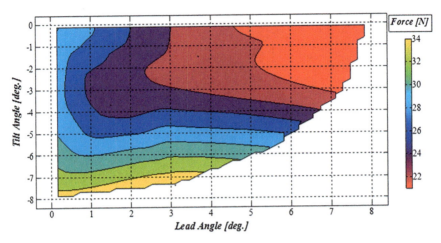

Fig. 3.13 Average of cutting force normal to machined surface for up milling operation

component force with respect to the machined surface change with respect to the lead and tilt angle, it would be necessary to find the optimum orientation in which lower cutting force is exerted on workpiece. Figures 3.12 and 3.13 show the contour plot in which the optimum tool orientation can be chosen. According to Fig. 3.12, the maximum average of normal cutting force occurs at zero lead and tilt angles. However, for the up milling case, the highest normal force happens at zero lead and higher negative values of tilt angle. The optimum lead and tilt angles are defined with red color in Figs. 3.12 and 3.13. Obviously, the selection of optimum lead and tilt angle should be based on the geometry of the process as well. There should always be checking criteria to make sure that the selected lead and tilt angle is feasible from geometrical point of view. This literally means to check for having collusion free machining operation.

3.4 Micro Ball End Milling of Titanium

3.4.1 Kinematics of the Process

Modeling of micro milling forces is a key factor to improve machining quality, control and understand kinematics of cutting process. From the prior research, it is now clear that micro milling is different from conventional machining operations and besides scaling down of the process presence of specific phenomena was found. The most frequent phenomenon faced during micro milling is the size effect. The size effect occurs because the edge radius of the micro tool is in the same range with uncut chip thickness. This phenomenon results in plowing of

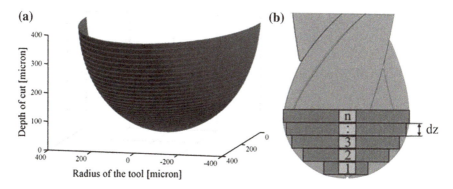

Fig. 3.14 **a** Discretization of ball end mill. **b** Schematic view of discretization

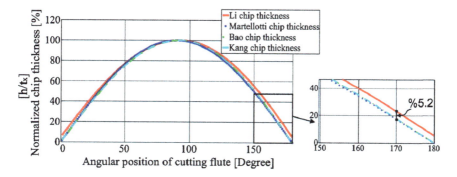

Fig. 3.15 Comparison of chip thickness models for the feed per tooth to tool radius of 0.1

workpiece material where no chip formation occurs because the chip thickness is less than minimum chip thickness which depends on material of machined part.

A new methodology for modeling cutting forces for ball end mill is proposed in this chapter. The ball end mill is discretized to finite number of end mills according to local radius as shown in Fig. 3.14 and forces calculated at each disk are integrated to calculated total cutting force in three Cartesian directions.

From the prior research, it is clear that chip thickness is the primary input for estimating cutting forces. The expression of the chip thickness for conventional milling was proposed by Martellotti [38], but this equation is not valid for micro milling, where the ratio of feed per tooth to tool radius is higher than in conventional milling. Several research groups developed chip thickness models for micro milling. In the Fig. 3.15 normalized chip thickness models proposed by Martellotti [38], Bao et al. [39], Kang et al. [40] and Li et al. [33] are compared.

During cutting force estimation it was seen that chip thickness model developed by Li has the best match with experimental results. The same chip thickness model is utilized in the developed force prediction model. The proposed chip thickness model is given below,

Fig. 3.16 Cutting force
components

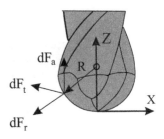

$$(h)_k = R \cdot \left[1 - \sqrt{1 - \frac{2\, t_x \sin\theta}{R + \frac{N t_x}{2\pi} \cos\theta} - \frac{t_x^2 \cos(2\theta)}{\left(R + \frac{N t_x}{2\pi}\cos\theta\right)^2} - \frac{t_x^3 \sin\theta \cos^2\theta}{\left(R + \frac{N t_x}{2\pi}\cos\theta\right)^3}} \right]$$

(3.11)

where $(h)_k$ is chip thickness, R is the local radius of the tool, N is number of flutes, t_x is feed per tooth and θ is rotational angle. Force model for predicting forces during the micro milling process with ball end mill was established using mentioned chip thickness model.

The instantaneous chip load for kth disk can be written as following:

$$dA_c = (h)_k \cdot (dz)_k$$

(3.12)

The differential cutting forces in tangential, radial and axial directions (t, r, ψ) are written as

$$\begin{bmatrix} dF_t \\ dF_r \\ dF_\psi \end{bmatrix} = \begin{bmatrix} K_{tc} dA_c \\ K_{rc} dA_c \\ K_{\psi c} dA_c \end{bmatrix} + \begin{bmatrix} K_{te} dz \\ K_{re} dz \\ K_{\psi e} dz \end{bmatrix}$$

(3.13)

where dF_t, dF_r and dF_ψ are differential forces in tangential, radial and axial direction shown in Fig. 3.16 and K_{tc}, K_{rc} and $K_{\psi c}$ are cutting force coefficients, K_{te}, K_{re} and $K_{\psi e}$ are cutting edge coefficients. Cutting force and edge coefficients vary along tool axis direction and are determined by calibration procedure described in Lazoglu et al. [41] paper.

$$T = \begin{bmatrix} \cos(\theta) & \sin(\theta) & 0 \\ -\sin(\theta) & \cos(\theta) & 0 \\ 0 & 0 & 1 \end{bmatrix}$$

(3.14)

By using transformation matrix T given in Eq. 3.14 tangential, radial and axial forces are transformed to forces in three Cartesian directions—X, Y and Z as following:

$$\begin{bmatrix} dF_x \\ dF_y \\ dF_z \end{bmatrix} = [T] \times \begin{bmatrix} dF_t \\ dF_r \\ dF_\psi \end{bmatrix}$$

(3.15)

Fig. 3.17 Experimental setup

Table 3.3 Cutting conditions and tool specifications

Tool diameter	800 (μm)
Number of flutes	2
Measured helix angle	25.2°
Spindle speed	12,000 rpm
Feed rate	120 mm/min
Depth of cut	100–400 (μm)

3.4.2 Simulation and Experimental Results

Experiments are performed on Ti-6Al-4V grade titanium alloy using 800 μm diameter two fluted Tungsten Carbide micro end mill on a 5-axis CNC milling machine and cutting forces are measured with table type mini dynamometer. The experimental setup used for validation of force model is shown below in Fig. 3.17.

Cutting experiments were performed on Ti-6Al-4V at 12,000 rpm spindle speed, at half immersion and slot cutting conditions, with the feed per tooth values of 5 microns at different depth of cut. Table 3.3 presents the cutting conditions and tool specifications used in the simulation and the experiments for micro milling. Using a calibrated tool corresponding cutting coefficients are estimated and given in Table 3.4.

Experimental forces for different cutting conditions were measured by mini dynamometer to validate the force model. Estimated and measured forces for slot milling condition are presented in Figs. 3.18 and 3.19.

Estimated and measured forces for half immersion cutting condition are presented in Figs. 3.20 and 3.21.

Table 3.4 Cutting coefficients for Ti-6A-4V with Tungsten Carbide cutting tool

Cutting disk number	Cutting force coefficients			Cutting edge coefficients		
	K_{tc} (N/mm^2)	K_{rc} (N/mm^2)	$K_{\psi c}$ (N/mm^2)	K_{te} (N/mm)	K_{re} (N/mm)	$K_{\psi e}$ (N/mm)
1	1080.6	540.2	650.9	3.071	1.04	3.567
2	397.1	234.7	213.5	0.806	0.618	1.099
3	286.2	139.3	155.9	0.531	0.461	0.415
4	158.1	95.5	88.1	0.398	0.247	0.203

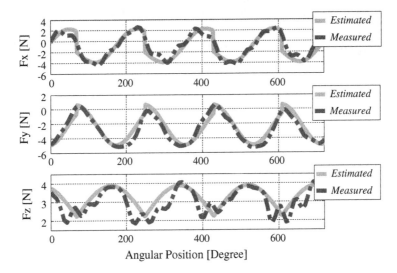

Fig. 3.18 Cutting forces for 100 μm depth of cut at slot milling cutting condition

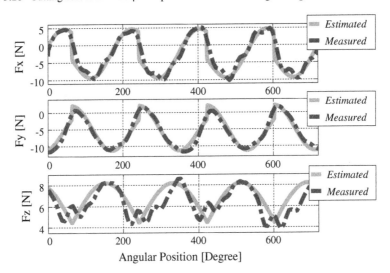

Fig. 3.19 Cutting forces for 300 μm depth of cut at slot milling cutting condition

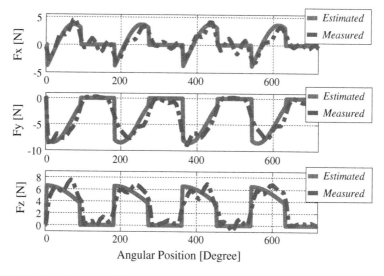

Fig. 3.20 Cutting forces for 200 μm depth of cut at half immersion cutting condition

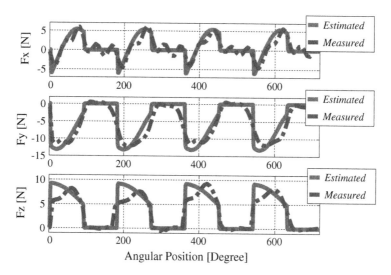

Fig. 3.21 Cutting forces for 400 μm depth of cut at half immersion cutting condition

3.5 Conclusion

In this chapter, the mechanics of ball end milling for macro and micro scale was investigated. It was shown that the orientation of the tool plays a critical role in mechanics of the five axis milling process. It is possible to optimize the tool orientation to minimize the cutting forces, which in turn will increase the integrity and quality of the finished surface. It is also possible to find an optimum condition in which the cutting torque and consumed energy is minimum.

Moreover, a new analytical model for predicting micro milling cutting forces for ball end mill were presented. The presented cutting force model is a function of cutting force coefficients and micro chip thickness model. The model was validated by experimental force measurements. Through the sets of the experiments, it was seen that the force model showed good agreement with experimental data.

References

1. Ezugwu EO (2005) Key improvements in the machining of difficult-to-cut aerospace superalloys. Int J Mach Tools Manuf 45(12–13):1353–1367
2. López de lacalle LN, Pérez J, Llorente JI, Sánchez JA (2000) Advanced cutting conditions for the milling of aeronautical alloys. J Mater Process Technol 100(1–3):1–11
3. Sun S, Brandt M, Dargusch MS (2009) Characteristics of cutting forces and chip formation in machining of Titanium alloys. Int J Mach Tools Manuf 49(7–8):561–568
4. Jaffery SI, Mativenga PT (2009) Assessment of the machinability of Ti-6Al-4V alloy using the wear map approach. Int J Adv Manuf Technol 40(7–8):687–696
5. Costes JP, Guillet Y, Poulachon G, Dessoly M (2007) Tool-life and wear mechanisms of CBN tools in machining of Inconel 718. Int J Mach Tools Manuf 47(7–8):1081–1087
6. Li HZ, Zeng H, Chen XQ (2006) An experimental study of tool wear and cutting force variation in the end milling of Inconel 718 with coated carbide inserts. J Mater Process Technol 180(1–3):296–304
7. Liao YS, Lin HM, Wang JH (2008) Behaviors of end milling Inconel 718 superalloy by cemented carbide tools. J Mater Process Technol 201(1–3):460–465
8. Rahman M, Seah WKH, Teo TT (1997) The machinability of Inconel 718. J Mater Process Technol 63(1):199–204
9. Devillez A, Schneider F, Dominiak S, Dudzinski D, Larrouquere D (2007) Cutting forces and wear in dry machining of Inconel 718 with coated carbide tools. Wear 262(7–8):931–942
10. Amin AKM, Ismail AF, Nor Khairusshima MK (2007) Effectiveness of uncoated WC–Co and PCD inserts in end milling of titanium alloy—Ti–6Al–4V. J Mater Process Technol 192:147–158
11. Arrazola P-J, Garay A, Iriarte L-M, Armendia M, Marya S, Le Maître F (2009) Machinability of titanium alloys (Ti6Al4V and Ti555.3). J Mater Process Technol 209(5):2223–2230
12. Fang N, Wu Q (2009) A comparative study of the cutting forces in high speed machining of Ti–6Al–4V and Inconel 718 with a round cutting edge tool. J Mater Process Technol 209(9):4385–4389
13. Pawade RS, Joshi SS, Brahmankar PK, Rahman M (2007) An investigation of cutting forces and surface damage in high-speed turning of Inconel 718. J Mater Process Technol 192–193:139–146
14. Kitagawa T, Kubo A, Maekawa K (1997) Temperature and wear of cutting tools in high-speed machining of Inconel 718 and Ti-6Al-6 V-2Sn. Wear 202(2):142–148

15. Thakur DG, Ramamoorthy B, Vijayaraghavan L (2009) Study on the machinability characteristics of superalloy Inconel 718 during high speed turning. Mater Des 30(5):1718–1725
16. MacGinley T, Monaghan J (2001) Modelling the orthogonal machining process using coated cemented carbide cutting tools. J Mater Process Technol 118(1–3):293–300
17. Cotterell M, Byrne G (2008) Dynamics of chip formation during orthogonal cuttingorthogonal cutting of titanium alloy Ti–6Al–4V. CIRP Ann Manuf Technol 57(1):93–96
18. Ng E-G, Lee DW, Sharman ARC, Dewes RC, Aspinwall DK, Vigneau J (2000) High speed ball nose end milling of Inconel 718. CIRP Ann Manuf Technol 49(1):41–46
19. Mhamdi M-B, Boujelbene M, Bayraktar E, Zghal A (2012) Surface integrity of Titanium alloy Ti-6Al-4V in ball end milling. Phys Procedia 25:355–362
20. Sonawane HA, Joshi SS (2012) Analysis of machined surface quality in a single-pass of ball-end milling on Inconel 718. J Manuf Process 14(3):257–268
21. Corduan N, Himbet T (2003) Wear mechanisms of new tool materials for TiBAl4V high perfonnance machining. CIRP Ann Manuf Technol 51(1):73–76
22. Aramcharoen A, Mativenga P, Yang S, Cooke K, Teer D (2008) Evaluation and selection of hard coatings for micro milling of hardened tool steel. Int J Mach Tools Manuf 48(14):1578–1584
23. Özel T, Sima M, Srivastava AK, Kaftanoglu B (2010) Investigations on the effects of multi-layered coated inserts in machining Ti–6Al–4V alloy with experiments and finite element simulations. CIRP Ann Manuf Technol 59(1):77–82
24. Thepsonthi T, Özel T (2013) Experimental and finite element simulation based investigations on micro-milling Ti-6Al-4V Titanium alloy: effects of cBN coating on tool wear. J Mater Process Technol 213(4):532–542
25. Vogler M, Kapoor S, DeVor R (2004) On the modeling and analysis of machining performance in micro-end milling, Part II: cutting force prediction. Trans ASME J Manuf Sci Eng 126(4):695–705
26. Waldorf DJ, DeVor R, Kapoor S (1998) Slip-Line field for ploughing during orthogonal cuttingorthogonal cutting. Trans ASME J Manuf Sci Eng 120(4):693–698
27. Jun MBG, Liu X, DeVor RE, Kapoor SG (2006) Investigation of the dynamics of microend milling—part I: model development. J Manuf Sci Eng 128(4):893
28. Fang N (2003) Slip-line modeling of machining with a rounded-edge tool—Part I: new model and theory. J Mech Phys Solids 51(4):715–742
29. Fang N, Jawahir IS (2002) An analytical predictive model and experimental validation for machining with grooved tools incorporating the effects of strains, strain-rates, and temperatures. CIRP Ann Manuf Technol 51(1):83–86
30. Rodríguez P, Labarga JE (2013) A new model for the prediction of cutting forces in micro-end-milling operations. J Mater Process Technol 213:261–268
31. Jin X, Altintas Y (2011) Slip-line field model of micro-cutting process with round tool edge effect. J Mater Process Technol 211(3):339–355
32. Park SS, Malekian M (2009) CIRP annals—manufacturing technology mechanistic modeling and accurate measurement of micro end milling forces. Micro 58:49–52
33. Li C, Lai X, Li H, Ni J (2007) Modeling of three-dimensional cutting forces in micro-end-milling. J Micromech Microeng 17(4):671–678
34. Altintas Y (2012) Manufacturing automation: metal cutting mechanics, machine tool vibrations, and CNC design. Cambridge university press, Cambridge
35. Merchant ME (1944) Basic mechanics of the metal cutting process. J Appl Mech 11(A):168–175
36. Layegh SEK, Erdim H, Lazoglu I (2012) Offline force control and feedrate scheduling for complex free form surfaces in 5-axis milling. Procedia CIRP 1:96–101
37. Gradišek J, Kalveram M, Weinert K (2004) Mechanistic identification of specific force coefficients for a general end mill. Int J Mach Tools Manuf 44(4):401–414
38. Martellotti M (1941) An analysis of milling process. Trans ASME J Manuf Sci Eng 63:677–700

78 I. Lazoglu et al.

39. Bao W (2000) Modeling micro-end-milling operations. Part I: analytical cutting force model. Int J Mach Tools Manuf 40(15):2155–2173
40. Kang YH, Zheng CM (2013) Mathematical modelling of chip thickness in micro-end-milling: a Fourier modelling. Appl Math Model 37(6):4208–4223
41. Lazoglu I, Boz Y, Erdim H (2011) Five-axis milling mechanics for complex free form surfaces. CIRP Ann Manuf Technol 60(1):117–120

Chapter 4
Analysis of Physical Cutting Mechanisms and Their Effects on the Tool Wear and Chip Formation Process When Machining Aeronautical Titanium Alloys: Ti-6Al-4V and Ti-55531

Mohammed Nouari and Hamid Makich

Abstract The current research deals with the analysis of physical cutting mechanisms involved during the machining process of titanium alloys: Ti-6Al-4V and Ti-55531. The objective is to understand the effect of all cutting parameters on the tool wear behavior and stability of the cutting process. The investigations have been focused on the mechanisms of chip formation and their interaction with tool wear. At the microstructure scale, the analysis confirms the intense deformation of the machined surface and shows a texture modification. As the cutting speed increases, cutting forces and temperature show different progressions depending on the considered microstructure Ti-6Al-4V or Ti-55531 alloy. Results show for both materials that the wear process is facilitated by the high cutting temperature and the generation of high stresses. The analysis at the chip-tool interface of friction and contact nature (sliding or sticking contact) shows that the machining Ti-55531 often exhibits an abrasion wear process on the tool surface, while the adhesion and diffusion modes followed by coating delamination process are the main wear modes when machining the usual Ti-6Al-4V alloy. Moreover, the proposed study describe the real effect on machining of the tool geometry, coating and lubrication. Finally, the investigations allow to identify some ways to improve the machinability of these alloys, particularly the Ti-55531 alloy.

M. Nouari (✉)
Laboratoire D'Énergétique et de Mécanique Théorique et Appliquée, LEMTA CNRS-UMR 7563, GIP-InSIC, 27 Rue D'Hellieule, 88100 St-Dié-Des-Vosges, France
e-mail: mohammed.nouari@univ-lorraine.fr

M. Nouari
University of Lorraine/Mines Nancy, Mines Nancy, France

H. Makich
Laboratoire D'Énergétique et de Mécanique Théorique et Appliquée, LEMTA CNRS-UMR 7563, GIP-InSIC/Mines D'Albi, 27 Rue D'Hellieule, 88100 St-Dié-Des-Vosges, France

J. P. Davim (ed.), *Machining of Titanium Alloys*,
Materials Forming, Machining and Tribology, DOI: 10.1007/978-3-662-43902-9_4,
© Springer-Verlag Berlin Heidelberg 2014

4.1 Introduction

Titanium alloys are widely used for applications requiring an excellent mechanical resistance and high strength at elevated temperature. This is the case of the Boeing 787 and A350 whose several structural parts are made in a new titanium alloy: the Ti-55531. This alloy has high mechanical characteristics that allow better performance on commercial aircraft, and a significant gain compared to the commonly used Ti-6Al-4V alloy. Beyond its interesting ratio density/mechanical properties, the Ti-55531 alloy provides significant benefits for the demanding environment of aeronautics. This comes from its ability to maintain mechanical properties at high temperature, the fatigue strength and its excellent corrosion resistance [1, 2]. Clément et al. [3, 4] have reported that the high-level properties of titanium alloys depend on several strengthening mechanisms such as grain size, solid solution atoms, and precipitation hardening, which all can be tuned during the various forming processing steps, leading to particular microstructures.

In manufacturing production, titanium alloys are classified as hard to cut materials. The main problems encountered when machining titanium alloy are the low material removal rate and the short tool life because of the excessive wear exhibited during the chip formation process. The first findings of machining Ti-55531 alloy are manifested by low cutting speeds. This causes high cycle times and reduced tool life, and then generates an increase in manufacturing costs.

Titanium is chemically reactive and, therefore, has a tendency to weld to the cutting tool during machining leading to chipping and premature tool failure [5–7]. In addition, its low thermal conductivity (about 15 W/m K vs. 270 W/m K for the steel CRS1018 at 700 °C) increases the temperature at the tool/workpiece interface, which adversely affects the tool life. Additionally, the high strength maintained at elevated temperature and low modulus of elasticity (50 % less than that of the steel) further impairs the machinability of these materials [6, 8, 9]. According to Ezugwu [10], the tool wear in machining of titanium alloys is due to high stresses and high temperatures found near to the cutting edge. The same conclusions were made by Subramanian [11].

From a microstructural point of view, elemental titanium presents an allotropic phase transformation at 880 °C between the body-centered cubic (bcc) and hexagonal- close-packed structures stable at high and low temperatures, respectively. The two phases of titanium alloys are known as α phase and β phase respectively [12]. Combinations of working and heat treatment alter the microstructure and change the mechanical properties of the metal. The microstructure and properties can also be affected by adding other elements to titanium. Addition of other elements to pure titanium, i.e. alloying, can alter microstructure and properties as well. Depending on which phase is to be dominant in a particular alloy (α, β or $\alpha + \beta$) an alloying element (or group of elements) may be added to pure titanium [3–5]. Thus, one way of classifying alloying elements is according to whether they are α or β stabilizers. Alpha stabilizers are soluble in the α-phase and many act as solid solution strengtheners while also increasing the temperature at which the

α-phase is stable. Alpha stabilizers include such elements as Al, Ga, Sn, Ge, and La. Beta stabilizing elements decrease the β transus (i.e., the temperature at which the material transforms to 100 % β-phase). As such, these elements increase the range over which the β-phase is stable. Beta stabilizers may be isomorphous or eutectoid. Isomorphous elements (such as V, Mo, Nb, Ta, and Re) are soluble in the α-phase while eutectoid elements like (Cr, Fe, Mn, Cu, Ag, Au, Ni, and Co) create a eutectoid phase.

According to Fanning [13], Ti-55531 (TIMETAL 555) is a high-strength near-β titanium alloy that was designed for improved productivity and excellent mechanical property combinations, including deep hardenability especially in aeronautical and aerospace industries. According to the same author and to Clément et al. in [3, 4], this recent alloy was designed based on the older Russian alloy VT22 to primarily fulfill high-strength forging applications. A lower-strength state with improved toughness and damage tolerance is under consideration for other parts of aircraft structure [14]. However, being of recent origin, there is a lack of information on machining of this alloy. This is despite the existence of information regarding thermomechanical processing of this family of alloys [15], and the abundance of information available regarding machining of titanium and some of its alloys (notably Ti-6Al-4V) [16]. In the machining field, it is well known that titanium and titanium alloys are hard-to-cut materials. This means that Ti-55531 will most likely present engineers with many technical problems to be solved in order to produce net shape components. Bouchnak recently proposed in [12] a study on the machinability of this material and assistance techniques to enhance the machining process. However, according to the literature in the field of machining, the state of the art does not present optimal solutions or does not give the real parameters that influence the cutting forces and tool wear (cutting conditions, tool geometry, and cutting material).

In order to increase productivity and tool-life in machining of titanium alloys, it is necessary to study the chip formation process and its effect on the physical cutting parameters and the material cutting performance. The objective of this study is to understand the poor machinability of titanium alloys especially the Ti-55531 one which exhibits extreme tool wear and unstable cutting forces.

4.2 Experimental Procedure

4.2.1 Workpiece Material

To analyse the machinability of the titanium alloys Ti-6Al-4V and Ti-55531, a parametric study has been conducted through several orthogonal cutting tests (cutting tests). Instrumented the Chip formation and tool wear processes have particularly been investigated in terms of cutting parameters such as cutting forces, friction coefficient and cutting temperature. Tests were carried out on a heavy-duty

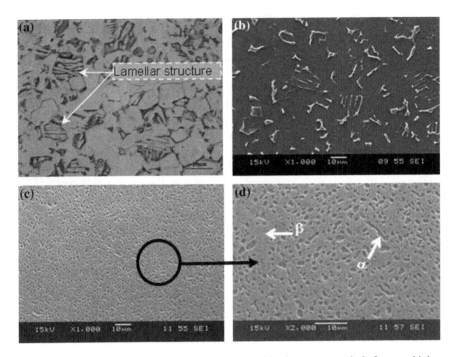

Fig. 4.1 Microstructure of the workpiece material used in the current study before machining. **a** Optical micrograph of Ti-6Al-4V ($\alpha + \beta$ alloy Ti-6Al-4V). **b** SEM micrograph of Ti-6Al-4V. **c** SEM micrograph of Ti-55531 (β alloy Ti-55531). **d** High magnification of SEM micrograph of Ti-55531

lathe machine with an 11 kW motor drive, which generates a maximum torque of 1,411 Nm. The spindle rotational speed ranges from 18 to 1800 rpm.

As mentioned above, two alloys were chosen for machining experiments. The first one is the alloy Ti-6Al-4V, considered as the reference of these tests. It is characterized with duplex structure $\alpha/\alpha+\beta$, and average grain size around 10 µm (range from 5 to 20 µm). The second alloy studied here is the Ti-55531 with average grain size around 1 µm (range from 0.5 to 5 µm).

Figure 4.1a–d depict the microstructure of each workpiece alloy before machining. The initial microstructure of the Ti-6Al-4V alloy, Fig. 4.1a, b, consists of single phase α matrix. Inclusions of β grains can also be seen with average grain size of 10 µm (range of 5–20 µm). Similarly, the initial microstructure of the β alloy Ti-55531 is shown in Fig. 4.1c, d. The microstructure consists of single phase β matrix with average size of α grains of 5 µm (range 1–5 µm).

In this work it has been found that the lamellar structure can be observed in $\alpha + \beta$ colonies (transformed β); particularly for the Ti-6Al-4V alloy. Figure 4.1a shows the localisation of the lamellar structure. This was previously confirmed by the work of Benedetti and Fontanari in [17].

Table 4.1 Chemical composition of machined titanium

Chemical element	Ti-6Al-4V (wt%)	Ti-55531 (wt%)
Al	5.5	5
V	3.8	5
Fe	Max 0.8	0.3
Mo	0	5
Cr	0	3
Nb	0	0.5–1.5
Zr	0	0.5–1.5

Table 4.2 Mechanical and thermal properties of Ti-6Al-4V and Ti-55531

	Ti-6Al-4V	Ti-55531
B transus (T_β) (°C)	980	856
Density (g/cm^3)	4.43	4.65
Tensile elastic modulus (GPa)	110	112
Compressive elastic modulus (GPa)	–	113
Tensile strength (MPa)	931	1236
Yield strength (MPa)	862	1174
Elongation (%)	14	6
Thermal conductivity at 20 °C (W/m K)	7.3	6.2
Specific heat 20–100 °C (J/Kg K)	709	495

Table 4.1 presents a summary of the chemical composition of both alloys. According to Fanning [13], Nouari et al. [18] and Bouchnak [12], the physical properties of the Ti-6Al-4V and Ti-55531 alloys are summarized in Table 4.2.

To complete the characterization of studied titanium alloys, tests of Vickers hardness have been performed on different specimens under room temperature. The micro-hardness of the Ti-6Al-4V specimen was found to be about $317HV_{0.2}$ and that measured for the Ti-55531 alloy to be about $379HV_{0.2}$. The Ti-55531 is therefore 20 % harder than Ti-6Al-4V. This result confirms the hard nature of the Ti-55531 microstructure, which will have a direct effect on its machinability, i.e. the level of cutting forces, tool wear, cutting temperature, etc.

4.2.2 Cutting Tools

4.2.2.1 Cutting Tool Material

In this study, inserts made of tungsten carbide (WC-Co) were used. The effect of the coating on the tool wear under extreme loading conditions was also investigated. To do that, a comparative study was performed between uncoated tool and coated one using a single layer of TiAlN coating with average thickness of 4 μm.

Fig. 4.2 SEM micrograph of cemented carbide tool (WC-6 %Co). **a** Tool microstructure. **b** Co and Wc distribution at the tool surface

Table 4.3 Mechanical and thermal properties of the cutting tool substrate [5, 6, 8]

Tool substrate	WC-6 %Co
Hardness 25 °C (HV_{10})	1,485
Hot hardness 800 °C (kg/mm^2)	600
Density (g/cm^3)	11.4
Thermal conductivity (W/mK)	45
Thermal expansion (10^{-6}/K)	6.1
Modulus of elasticity (GPa)	620
Traverse rupture (GPa)	2.2
Poisson coefficient υ	0.26

The TiAlN coating has a strong chemical stability, a low thermal conductivity and an high oxidation wear resistance at 900 °C. The TiAlN coating increases the surface hardness to approximately 3400–3600 Hv and improves the resistance to abrasive wear. The thermal conductivity this coating is about 10 W/mK (at 20 °C). Devillez et al. [19] state that TiAlN coating imparts an excellent crater resistance. Additionally, Singh et al. [20] and Castanho et al. [21] show that the Al element incorporated in TiAlN coating forms the superficial layer Al_2O_3 to improve the wear resistance and to enhance the chemical stability.

Therefore, the cutting tools used in these machining tests are made of tungsten carbide with cobalt binder, Grade H13A (WC-6 %Co, type K20). The tool microstructure is shown in Fig. 4.2a. Average and maximum sizes of the WC grains are respectively 1–5 μm. The average percentage of cobalt is about 6 %, the analysis under a scanning electron microscope (SEM) revealed that the Co binder is uniformly distributed in the tool surface, Fig. 4.2b. Table 4.3 presents a summary of the mechanical and thermal properties of cutting tools.

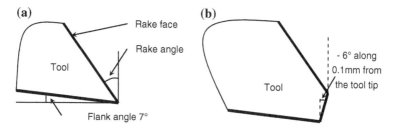

Fig. 4.3 Presentation of the tool geometries used in machining tests. **a** *Tool A, Rake angle* constant along the *rake face*. **b** *Tool B*, two different rake angles along the cutting edge

4.2.2.2 Cutting Tool Geometries

Concerning the tool geometry, a special designed tool (with two different geometries) has been considered in these experiments. Figure 4.3 shows the two geometries used in the study. The first geometry is more conventional without a particular treatment on the cutting edge, designated by geometry "A" (Fig. 4.3a). The second geometry is characterized by an additional angle of the cutting edge, designated by geometry "B" (Fig. 4.3b).

Tools for orthogonal cutting (width 4.45 mm) have one cutting edge and the flank angle is about 7°. Also, the tool "A" has a rake angle of 20° and the tool "B" has a rake angle of 0°. The tool geometries were tested for both materials in the study under orthogonal machining configuration.

4.2.3 Cutting Parameters

To determine the lubrication influence on the tool wear when machining titanium alloys, experimental tests were performed with and without lubrication. These investigations are added to the analysis of the influence of the material microstructure, tool geometry and coating on machinability titanium alloys. However, machining is a manufacturing process with a large number of interacting variables. The produced geometry is influenced by many variables, such as cutting speed, feed, depth of cut, etc. Therefore, the cutting conditions of Table 4.4 are taken into account to achieve a parametric study on orthogonal cutting.

Thus, all these parameters (cutting conditions, tool coating and geometry and lubrication) were considered as wear factors in this work to investigate the influence of cutting forces, friction coefficient, cutting temperature, chip formation process on tool wear.

As shown by Table 4.4, experiments were carried out keeping cutting speed and rake angle at various levels. The range of each factor was selected based on the present day industrial requirements. The cutting length allowed by the machine

Table 4.4 Cutting conditions and tool geometry

Cutting speed V_c (m/min)	20	35	65
Feed f (mm/rev)	0, 1	0, 1	0, 1
Rake angle (°)	0 and 20°	0 and 20°	0 and 20°
Flank angle (°)	7°	7°	7°

Fig. 4.4 Cutting forces and their direction in the machining plane. (F_c: Cutting force, F_a: Feed force, F_t: Tangential force, F_n: Normal force)

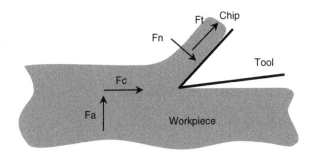

capacity (about 1.5 m) provides a sufficient cutting time to reach a stationary regime of the cutting process (1.6 s for a cutting speed of 65 m/min).

All experiments were carried out with a constant width of 3.5 mm. The other variables such as machine condition, variability in set up, etc. have been maintained constant throughout the experimentation. A three component Kistler® dynamometer was employed for cutting force measurements (Fig. 4.4). The forces reported are those for the process in a stable state with almost steady pulses.

4.2.4 Thermal Characterization of Titanium Alloys

4.2.4.1 Temperature Measurements

The cutting temperature was estimated using two techniques: the first technique is based on measurements with infrared camera (*Cedip camera*); and the second technique is based on an inverse measurement method using thermistor located behind the cutting tool edge. The last technique was previously developed by Battaglia et al. in [22].

Infrared camera provides directly the thermal field which only gives an assessment of temperature levels. This technique allows a qualitative analysis, unfortunately it cannot be used to determine with high precision temperature values on the cutting tool surface. This is due on one hand to the complexity of the instrumentation, and on the other hand to the confinement of the contact area between tool, chip and workpiece. Table 4.5 presents mean values of the cutting temperature in the case of machining Ti-55531. These measurements were performed using two different cutting speeds (V = 20 m/min and V = 65 m/min).

Table 4.5 Mean temperature obtained by infrared camera technique (Cedip camera) when machining Ti-55531 with WC-6 %Co tool

Cutting speed (m/min)	Temperature (°C)
20	628
65	761

Table 4.6 Estimated temperatures during machining of Ti-55531 using thermistors

Cutting speed (m/min)	Temperature (°C)
20	450
65	800

In contrast with the first technique, the second method uses thermistors and provides an indirect measurement of the cutting temperature. The thermistor (very small thermocouple with diameter d = 470 μm) is placed in a hole made by electroerosion process inside the tool without embrittlement of its integrity. Thermistors are held fixed close to the cutting tool face. This makes possible to obtain measurements of the heat flux transmitted to the cutting tool. Using the inverse model, developed by Battaglia et al. in [22], an average temperature can be obtained with the measured heat flux taken as an input parameter of the model. We recall here that this type of measurement can not be performed in the case of lubricated machining to avoid disturbance of the thermal field.

Table 4.6 shows the average temperatures estimated when machining Ti-55531 with two cutting speeds (V = 20 m/min and V = 65 m/min).

As a conclusion on the temperature measurements study, it can be said that whatever the used cutting speed, no phase transition should be expected, since the corresponding cutting temperature remains less than the β transus of titanium alloys which is about 880 °C.

Therefore no phase transition happens in the microstructure, confirming the assumption of Puerta Velasquez et al. in [23]. The later reported that only a severe shearing occurs in the titanium material during machining.

4.2.4.2 Thermal Conductivity Measurements

Before starting the analysis, it very important to remind the evolution of thermal conductivities for the considered titanium alloys with temperature (Fig. 4.5). This has directly an impact on the evolution of tool wear. The data used for plotting Fig. 4.5, were obtained by measuring thermal conductivities using "Hot Disk" method for several specimens under different temperatures [24]. The high temperatures were close to those obtained during machining the same materials.

Ti-55531 and Ti-6Al-4V alloys show a clear difference of about 30 % in their thermal conductivities at 700 °C (Fig. 4.5). At high temperatures, the Ti-55531 alloy conducts heat better than the Ti-6Al-4V alloy. This result is very interesting since it

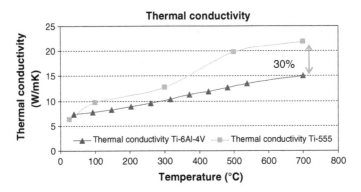

Fig. 4.5 Evolution of thermal conductivities of Ti-55531 and Ti-6Al-4V with cutting temperature

allows us to explain the morphology observed of formed chips and the variation in cutting forces. Indeed, a low thermal conductivity promotes chip segmentation and thus a reduction of the machining efforts; this is the case for Ti-6Al-4V alloy [25]. When the thermal conductivity increases, the heat stored in the machined material is discharged to the chip and then the ability of the machined material to soften is reduced. Under these conditions, the machining efforts will not fall because of the material flow stress which remains high even at very high temperatures, as in the case for the Ti-55531 alloy. Based on these findings, the degradation modes of the cutting tool could be of a mechanical type for Ti-55531 (abrasion mode) and physico-chemical for Ti-6Al-4V (adhesion and diffusion modes).

4.2.5 Monitoring of the Chip Formation Process

The study of the formation and the nature of chips is one of parameters used to characterize the machining. Due to the fact that machining process is very fast even at low cutting speeds, a high-speed camera (CCD) is used for viewing and monitoring chip formation (Fig. 4.6). The camera used in this study is Phantom®. The maximum resolution is 512 × 512. The acquisition in terms of frames number per second up to 11,000 frames/s. The analysis of different sequences give important information about the physical parameters of machining: worn contact length, deformation level, chips morphology, etc.

4.3 Results and Discussion

The obtained results are shown depending on whether one considers the lubricated or dry machining, the use of tools with or without coating. The calculation of the friction coefficient in various cases was also performed. The effects of the tool

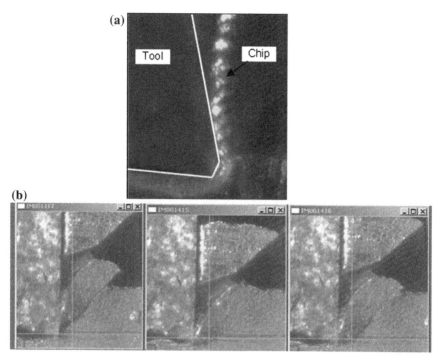

Fig. 4.6 Chip formation; **a** morphology, $V_c = 20$ m/min, rake angle $= 0°$, **b** segmentation process, $V_c = 65$ m/min, rake angle $= 0°$

geometry and the cutting speed on machinability are analyzed. Also an attempt has been made to examine the effect of various process parameters on the chip morphology and tool wear during machining two different titanium microstructures: Ti-6Al-4V and Ti-55531.

4.3.1 Effect of the Tool Geometry

The effect of the tool geometry on the machinability of titanium alloys Ti-6Al-4V and Ti-55531 is presented here with results on both alloys machined with TiAlN coated tools. The analysis of these results has been performed with a comparison on the machinability of the alloys according to the cutting forces, friction and the specific energy of the machining.

4.3.1.1 Effect on Cutting Forces

The cutting forces sign machining quality depending on cutting conditions, tool and workpiece materials. Thus, the measurements of forces give indications on

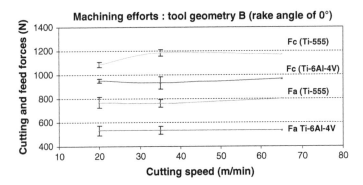

Fig. 4.7 Cutting forces obtained for Ti-6Al-4V and Ti-55531 with tool geometry "B"

machinability of the machined material. In addition to the friction at the tool/chip interface, cutting forces depend on two main factors: area of the primary and secondary shear planes, and shear strength of the work material at these planes [26].

Figures 4.7 and 4.8 illustrate cutting forces (F_c) and feed forces (F_a) generated during machining titanium alloys Ti-6Al-4V and Ti-55531. The evolution of machining efforts is plotted as a function of the cutting speed V_c and the considered tool geometry.

It can be observed from Figs. 4.7 and 4.8 that F_c is the dominant force component. Therefore, the discussion on cutting forces is focused on the cutting force due the weak variation of feed forces. As seen in Figs. 4.7 and 4.8, the feed force remains stable for all conditions.

First of all, it can be noted from Fig. 4.7, a stabilization of cutting forces with high cutting speeds (35 m/min and 65 m/min). However, a slight increase is still noticeable on the cutting force of the Ti-55531 alloy when cutting speed increases from 20 to 65 m/min. During machining the Ti-6Al-4V alloy with a tool rake angle of 20° (tool geometry "A") (Fig. 4.8), a reduction of cutting forces is noticed when the cutting speed increases. This is quite normal considering the thermal softening of the material due to the temperature rise during machining. In other words, when the cutting speed increases, the temperature increases too and this is followed by a decrease in the yield stress level of the alloy. The material deforms under these conditions much more easily and without significant effort. This trend is expected because machining becomes more adiabatic and the heat generated in the shear zone can not be conducted away during the very short interval of time during which the material passes through this zone. So, the temperature rise softens the material aiding grain boundary dislocation and thus reducing cutting forces as seen from Fig. 4.8.

However, the cutting speed can not be increased significantly pretext to further reduce the effort because of the significant increase (simultaneously) of the tool wear. On the other hand, an increase in the cutting force level of Ti-55531 with 20° tool rake angle can be observed when increasing the cutting speed. This tendency can be explained by the high strain rate sensitivity of the Ti-55531 alloy.

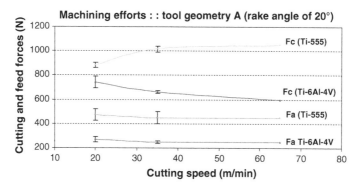

Fig. 4.8 Cutting forces obtained for Ti-6Al-4V and Ti-55531 with tool geometry "A"

According to Fig. 4.7 the comparison between generated cutting forces shows a clear difference between Ti-6Al-4V and Ti-55531. This difference is ranged from 12.3 to 21.1 % for the cutting force and from 30.1 to 33.4 % for the feed force. This result confirms the poor machinability (previously announced) of the Ti-55531 alloy compared to that of the Ti-6Al-4V under the same machining conditions. The trend given by tools with a rake angle of 0° is confirmed by tools with a rake angle of 20°. The difference may reach 42.9 % for some cutting conditions ($V = 65$ m/min and $\alpha = 20°$) (see Fig. 4.8).

Also, it was found from Figs. 4.7 and 4.8 that the cutting force obtained with the rake angle of 20° was weak compared to that of machining with 0° rake angle at all levels of considered parameters. This has been attributed to the fact that high values of the rake angle may reduce the friction along the cutting edge between tool and workpiece; see the evolution of the friction coefficient in Sect. 4.3.1.2).

4.3.1.2 Effect on Friction at the Tool/Chip Interface

In this section we focus on the friction, which is the manifestation of the mechanical energy dissipated in the contact between the tool and the workpiece in the form of heat which is responsible for the heating of the cutting edges.

Consequently, to understand physical phenomena during the chip formation, friction process have to be investigated at the tool-chip interface. The friction coefficient is an important parameter to characterize the nature of the tool-chip contact and tool wear depending on cutting conditions. According to the famous model of Merchant [16, 27], the calculation of this parameter can be obtained using cutting forces measurements. In machining, the apparent friction coefficient (or average friction) is often defined as the ratio between the tangential force F_t and the normal force F_n [16, 27].

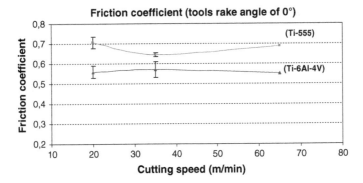

Fig. 4.9 Evolution of the friction coefficient in the tool/chip interface (Tool geometry "B", rake angle 0°)

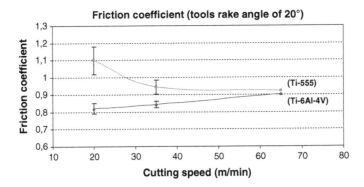

Fig. 4.10 Evolution of the friction coefficient in the tool/chip interface (Tool geometry "A", rake angle 20°)

$$\mu = \frac{F_t}{F_n} = \frac{F_a + F_c \tan \alpha}{F_c - F_a \tan \alpha} \qquad (4.1)$$

where α is the rake angle, F_c: cutting force, F_a: feed force, F_t: friction force, F_n: normal force.

Figures 4.9 and 4.10 show the variation of friction coefficient values at various cutting conditions. These values are calculated using Eq. (4.1). The results of this study show that friction depends on the materials, the cutting temperature and machining conditions (cutting speed, tool geometry, etc.). In all tests, this parameter is more important with tool geometry "A" (rake angle 20°) compared to those with 0° rake angle (tool geometry "B"). This is due to the reduction of the cutting temperature and contact length. This reduction is often followed by a reduction in frictional forces at the tool-workpiece interface. Low cutting temperatures reduce adhesion tendency of the cutting tool and promote contact area restriction. Reduction of the tool–chip contact length is expected to occur, promotion of the plastic flow at the backside of the chip and overall reduction of temperature.

The results presented above (in Figs. 4.9 and 4.10) show a higher friction coefficient for Ti-55531 compared to Ti-6Al-4V. Thus, the following assumptions can be presented: the contact area at the tool-chip interface can be considered as a sticking contact-type for Ti-55531 in the case of machining with tools of 0° rake angle. Therefore, a connection can be made between the nature of the contact (sticking) and the tendency of Ti-55531 to adhere to the cutting face of the tool. In contrast, the tool-chip contact can be of a sliding type in the case of Ti-6Al-4V because of the low values of friction (see Fig. 4.9). In the case of machining with tools of 20° rake angle (Fig. 4.10), the reduction in friction coefficient for the Ti-55531 indicate contact of sliding type more then sticking one. Contrary to this trend, increasing friction for the Ti-6Al-4V alloy indicate sticking contact. Also, the increase of the cutting speed (in particular for testing tool geometry 20°) involves the stabilization of the friction at high cutting speeds (the friction is a decreasing function of the temperature when the cutting speed increases).

In conclusion, at constant cutting speed, titanium alloy Ti-55531 is more sensitive to adhesion phenomena as Ti-6Al-4V alloy. This may also explain the poor machinability noticed on this alloy.

4.3.1.3 Effect on the Specific Energy in Machining

The specific energy in machining is also a good indicator of machinability of materials. It is calculated based on the cutting conditions and measured forces (Eq. 4.2).

$$u(J/m^3) = \frac{P}{Q} \tag{4.2}$$

with:

$$P = F_c \times V_c \tag{4.3}$$

where P is the cutting power and Q is the volume of removed material.

Figures 4.11 and 4.12 confirm the previous analysis of the difference between the machinability of Ti-6Al-4V and Ti-55531. The specific cutting energy is higher for the titanium alloy Ti-55531 to that of Ti-6Al-4V alloy.

It is also interesting to note here the effect of the tool geometry on the materials machinability. With a rake angle of 0° (tool "B"), the difference is around 12 % and remains constant regardless of the cutting speed. With a rake angle of 20° (tool "A"), the difference is greater and reaches high levels for the high cutting speeds [around 51 % for $V_c = 60$ m/min)]

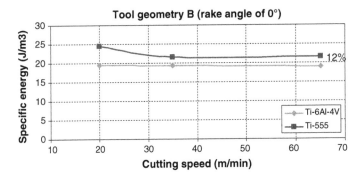

Fig. 4.11 Specific cutting energy for Ti-6Al-4V and Ti-55531, tool geometry "B"

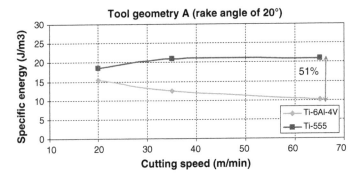

Fig. 4.12 Specific cutting energy for Ti-6Al-4V and Ti-55531, tool geometry "A"

4.3.2 Effect of Coating on Machinability

Additional tests were conducted on the Ti-55531 alloy to examine the effect of the coating tools on machinability. This is achieved without lubrication to prevent coupling problems. Only the tool geometry "B" with a rake angle of 0° was used. Thereby the tests are carried out under dry machining configuration with uncoated and coated tools with a single layer of TiAlN with average thickness of 4 μm.

Figures 4.13, 4.14 and 4.15 present the evolution of the machining efforts and friction at the tool/chip interface according to two cutting speeds 20 and 65 m/min.

It can be noted that for the same cutting conditions, the cutting forces obtained with the coated tool when machining Ti-55531 alloy are smaller than those obtained with the uncoated tools. The difference between the two cases is about 8 % for a cutting speed of 20 m/min. This difference is further increased when the speed increases, it reaches 24 % for 65 m/min. This shows that the coating improves significantly the machinability of the material. We can also observe from these results that the coated tools allow a reduction of efforts when the cutting speed increases for Ti-55531 alloy. The difference between the coated and

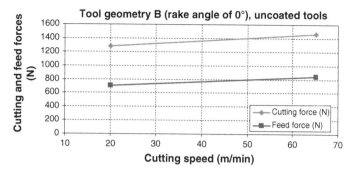

Fig. 4.13 Forces obtained for the Ti-55531 with tool geometry 0°, uncoated tool

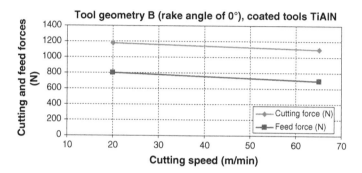

Fig. 4.14 Forces obtained for the Ti-55531 with tool geometry 0°, coated tool

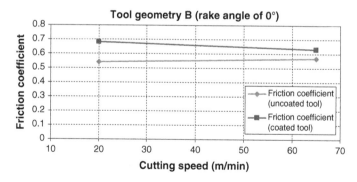

Fig. 4.15 Evolution of the friction coefficient in the tool/chip interface

uncoated tools is explained by the fact that the coating layer, taking into account the thermal and physicochemical characteristics amply improves the tool/chip contact. Its role as a thermal barrier allows the tool to protect themselves against the frictional heating and workpiece to soften more and to deform more easily.

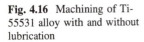
Fig. 4.16 Machining of Ti-55531 alloy with and without lubrication

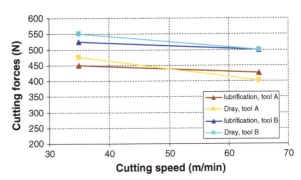

This interpretation can be confirmed by the analysis of friction at the tool/chip interface. Coated tools generated an intense friction (sticky type, μ close to 1). This means a more intimate contact between the tool and the workpiece. So, heat dissipation into the chips easier. However, the consequence of this contact is the ability to more easily form the bonding layers on the tool surface.

In the case of uncoated tools, friction is lower (0.55–0.57). Such a friction-type promotes a sliding contact and abrasive wear of the tool cutting face. The sliding contact does not facilitate the transmission of heat between the material and the tool, thus the alloy softening becomes difficult. This explains the increase in efforts to form a chip by an uncoated tool relative to a coated tool.

In conclusion, the coating significantly affects the machinability of Ti-55531 by the decrease of the machining efforts and the protection of the tool cutting edge.

4.3.3 Effect of Lubrication on Machinability

To study the effect of the lubrication, several tests were carried. The test results are presented on Fig. 4.16. The data on the Ti-55531 alloy does not clearly show the influence of the lubrication on the cutting forces and hence machinability. However, it has a major effect on tool wear (coated or not). Thus, the analysis shows that lubrication does not directly affect the machinability of Ti-55531 but indirectly via the tool wear. This conclusion differs from the effect of the tool coating; it has a direct effect on machinability and indirect effect on wear.

4.3.4 Review of Chips

In this section we present the results of analyzes performed on the chips. Two types of analysis can be discussed here: a microstructural analysis of chips and monitoring their formation with a high-speed camera (CCD).

Fig. 4.17 Chip formation
during machining

4.3.4.1 Chip Morphology of Titanium Alloys Ti-6Al-4V and Ti-55531

During machining and under the action of the tool cutting edge, the workpiece material undergoes a strong compression and deforms plastically. An intense shear is generated between the tool tip engaged in the material and the workpiece (Fig. 4.17). We have analyzed in our investigations this area, called "primary shear zone", in which occurs the first shear and induces the chip formation.

The chip morphology gives crucial information because it incorporates the material response during machining (mechanical, thermal, thermo-viscoplastic, etc.) and shows the stability of the cutting operation. Moreover, the evolution of cutting forces shows a correlation with the morphology of produced chips. The chip segmentation for example, often leads to a reduction of cutting efforts and the tool-chip contact length. In general, the continuous chip is undesirable in industrial applications. This is due to the generated problems and which significantly affect the machining conditions: damage of the cutting edge, congestion of the cutting area, etc.

The images obtained by the high speed camera (Fig. 4.18a) during machining of Ti-6Al-4V show a segmented chip. To confirm this observation further investigation were carried out on chips collected after machining. The micrographs obtained on these chips show a fairly regular segmentation (Fig. 4.18b). Between two consecutive segments of a single chip, the plastic deformation is very intense and localized in a thin zone of a few micrometers (Fig. 4.19). This area where the plastic deformation is localized is known as adiabatic shear band; it is the seat of extreme shear. The thickness of the shear band was measured for different cutting conditions and its evolution was followed depending on the material nature and on the value of cutting speed. The conclusion of this analysis is that the increase of cutting speed means that the shear bands are more refined in the chip body (Fig. 4.19).

Unlike Ti-6Al-4V, Ti-55531 chips have a morphology slightly scalloped with a rather continuous nature. They are also characterized by highly irregular thicknesses. This morphology retains the same characteristics at different cutting speeds and for the different geometries tested (Fig. 4.20).

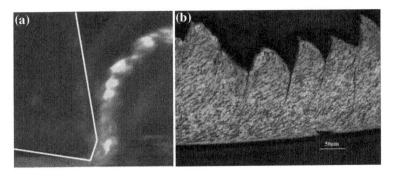

Fig. 4.18 Ti-6Al-4V alloy chip. **a** Chip formation (Ti-6Al-4V, 20 m/min, rake angle 0° (by high speed camera), **b** Chip morphology (Ti-6Al-4V, 35 m/min, rake angle 0°)

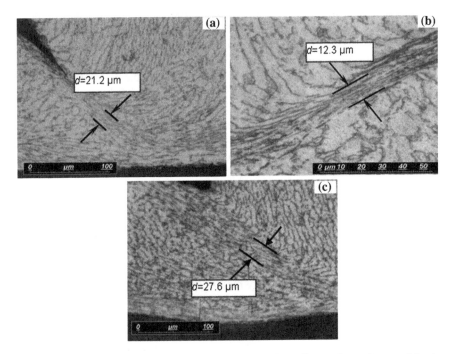

Fig. 4.19 Evolution of the adiabatic shear band thickness depending on the cutting speed for the Ti-6Al-4V. **a** 20 m/min, **b** 35 m/min, **c** 65 m/min. Tests without lubrication

Figure 4.21 shows the appearance of pronounced adiabatic shear bands without cracking or separation between the chip segments. So, the segmentation being synonyms of lower cutting efforts, during machining Ti-55531 alloy cutting efforts will not decreases even if cutting speed increases. This explains the poor machinability of Ti-55531 compared to the Ti-6Al-4V alloy.

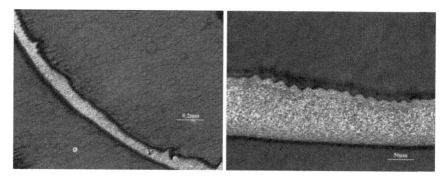

Fig. 4.20 Chip morphology of Ti-55531 (15 m/min, tool A)

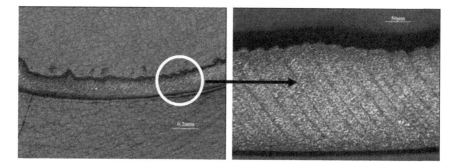

Fig. 4.21 Chip morphology of Ti-55531 (30 m/min, tool A)

Finally, under the same machining conditions and tool geometries, it can be seen that Ti-6Al-4V forms segmented chips more easily than the Ti-55531. This segmentation allows improved machinability resulted in lower cutting forces.

4.3.4.2 Chip Microstructure of Titanium Alloys Ti-6Al-4V and Ti-55531

Chips obtained after machining and presented in Fig. 4.22 were mounted with epoxy so that they stood on their edge in order to make the cross-section after polishing straight across its length. The polished chips and as-received workpiece material were etched with Kroll's reagent to reveal their microstructures. Micrographs of examined chips in Fig. 4.22 show clearly the deformation phenomenon inside the microstructure of both materials during the chip formation. However, the deformation process is different from one material to the other. It can be noted here that both chips given by machining Ti-6Al-4V and Ti-55531 alloys were obtained with the same cutting conditions.

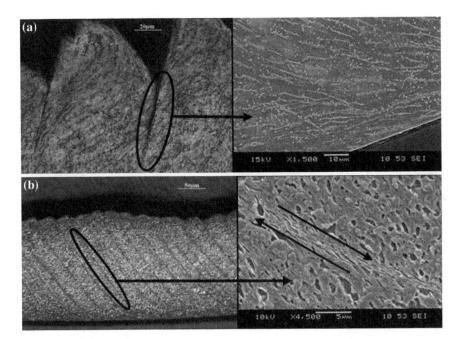

Fig. 4.22 Microstructure of the chip material obtained after machining. **a** Chip formation and high deformation of β phase for Ti-6Al-4V. **b** Chip formation and very localized deformation of β phase for Ti-55531

The examination of the deformed microstructure reveals very fine sizes of grains in the β alloy (Ti-55531). In this material, the deformation process is localized in a very thin layer called primary shear zone while in the Ti-6Al-4V alloy the same process of deformation occurs in the whole microstructure, i.e. in α, β and $\alpha + \beta$ phases. Indeed, chips of the Ti-6Al-4V alloy have a very different microstructure compared to the initial Ti-6Al-4V alloy. For Ti-55531 chips, apart from the primary shear zone, there is a very similar microstructure to that observed in the initial Ti-55531 alloy.

4.3.5 Wear Mechanisms of Cutting Tools

In this section the different modes of tools degradation have been analysed when machining titanium alloys Ti-6Al-4V and Ti-55531. Wear can be discussed in relation to the nature of the material involved, its metallurgical and thermophysical characteristics, machining conditions and the tool coating. The wear analysis is based on observations of the scanning electron microscope (SEM) equipped with energy X-ray spectrometer (EDS) in correlation with the measurements obtained

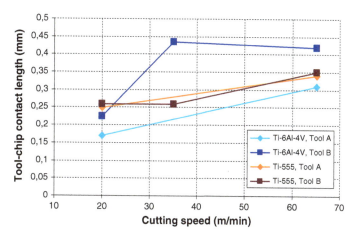

Fig. 4.23 Evolution of the tool-chip worn contact length

with an optical profilometer. The damage modes of tools characterization was conducted through the measurement of:

- The worn tool-chip contact length,
- The thickness and extent of the formed adhesion layers,
- The depth and width of the crater formed on the tool rake face.

4.3.5.1 Study of the Worn Tool-Chip Contact Length

The determination of the worn contact length was performed by analyzing the SEM micrographs of tool-chip contact zones. These areas have previously been identified by measuring the distance between the cutting edge and the limit of the worn portion on the interface. Furthermore, in situ analysis (using a high speed camera CCD) has shown that the chip winding is forwardly of the tool. Thus, the footprint of the wear on the tool surface represents the contact between the latter and the chip being formed. Figure 4.23 summarizes the different values of the recorded worn contact length.

The results show that the increase of the cutting speed induces an increase of the worn contact length. In the case of Ti-6Al-4V the variation is more remarkable. For both tested geometries, it is around 45 %. This confirms the analysis of the heat-softening behavior of material. Indeed, and as mentioned above, the Ti-6Al-4V has a lower thermal conductivity compared to Ti-55531 to the high temperatures (around 30 %). The stored heat thus generates greater softening for the Ti-6Al-4V than for the Ti-55531. The contact between the tool and the Ti-6Al-4V alloy then extends over a larger area of the tool surface relative to the Ti-55531 alloy. For the Ti-55531 alloy, the variation of the contact area is about 25 % when the cutting speed increases from 20 to 65 m/min. The presence of the

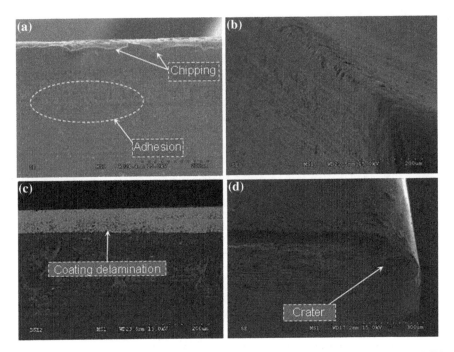

M. Nouari and H. Makich

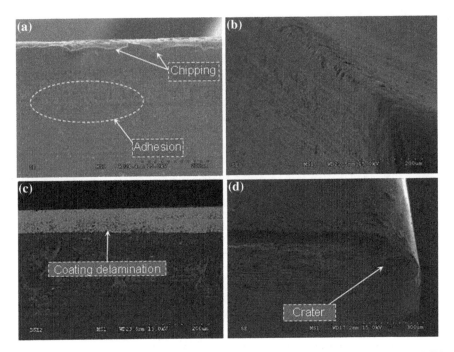

Fig. 4.24 SEM images (BSE and SE micrographs) of the tool wear when machining Ti-6Al-4V and Ti-55531 alloys with different cutting speeds (rake angle 0°). **a** SE micrograph (secondary electrons) for cutting tool (machining Ti-6Al-4V), V_c = 20 m/min. **b** SE micrograph (secondary electrons) for cutting tool (machining Ti-55531), V_c = 20 m/min. **c** BSE micrograph (Backscattered electrons) for cutting tool (machining Ti-6Al-4V), V_c = 65 m/min. **d** SE micrograph (secondary electrons) for cutting tool (machining Ti-55531), V_c = 65 m/min

betagenic element chromium in the Ti-55531 microstructure improves the deformation resistance of this alloy at high temperatures. Therefore, the Ti-55531 machinability is degraded as a function of temperature.

In conclusion, it can be said that the affected area by wear in the case of Ti-55531 is greater than in the case of Ti-6Al-4V at identical cutting conditions and tool geometry.

4.3.5.2 Tool Wear Analyses

It has been observed from the SEM analysis in Figs. 4.24 and 4.25 that the cutting tool encounters severe thermal and mechanical loading when machining titanium alloys. This can be supported by the level of the measured cutting temperature (about 750–800 °C for V = 65 m/min) and high recorded cutting forces (about 1,000 N). Also, other works previously showed that the cutting pressure can also attain large values (about 1–1.5 GPa) [10, 28]. The high stresses and high

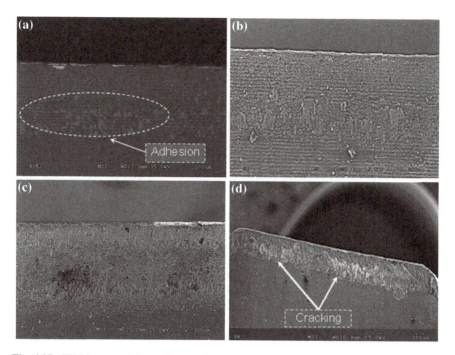

Fig. 4.25 SEM images of the tool wear when machining Ti-6Al-4V and Ti-55531 alloys (rake angle 20°). **a** Cutting tool (machining Ti-6Al-4V), V_c = 20 m/min. **b** Cutting tool (machining Ti-55531), V_c = 20 m/min. **c** Cutting tool (machining Ti-6Al-4V), V_c = 65 m/min. **d** Cutting tool (machining Ti-55531), V_c = 65 m/min

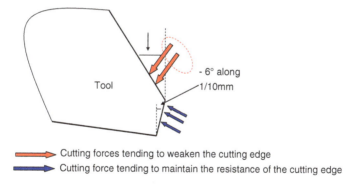

Fig. 4.26 Orientation of machining efforts on a tool edge

temperatures generated close to the cutting edge have great influence on the tool wear rate and on tool life.

The tool geometry "B" has been designed primarily to give a negative rake angle to the front of the cutting edge (about 0.1 mm). This allows steering the cutting forces inwardly of the tool (Fig. 4.26). The aim of this treatment undergone

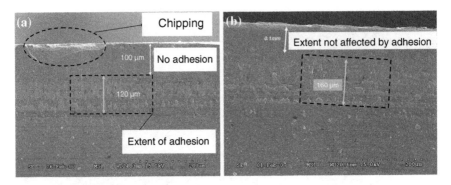

Fig. 4.27 Adhesion extent on tool geometry B (V_c = 15 m/min), **a** Ti-6Al-4V, **b** Ti-55531

by the cutting edge is to enhance the tool wear resistance. However, the test results showed a significant deterioration of machinability in terms of cutting forces compared to the tool geometry "A".

From Fig. 4.27, at identical machining conditions, it can be noted that the extent of the negative rake angle of 0.1 mm is not degraded by the adhesion process but only by abrasion wear. The most tangible explanation to this observation is that this area of the tool displaces (or rejects) the material rather than machined.

In the case of tools with 0° rake angle, the extent of the area affected by the wear is greater for Ti-55531 than for Ti-6Al-4V. At 20 m/min for example, it is about 120 μm for the Ti-6Al-4V while about 160 μm for Ti-55531 (Fig. 4.24a, b). This is mainly due to the fact that the increase in cutting forces for the Ti-55531 generates higher stress on the cutting edge. The hardness of the material also contributes to the heavy wear edge. Micrographs also show delamination of the coating layer on the tool surface during machining Ti-6Al-4V at 65 m/min. This is not the case for the Ti-55531 alloy under the same cutting speed (Fig. 4.24c, d).

The delamination phenomenon is difficult to analyze because it can have both thermal and mechanical origins. This is partly due to the complex interaction between various physical factors that control delamination: intrinsic properties of the coating, tool and those of the interaction between the substrate, coating and workpiece. However, it is possible to attribute the origin of delamination to chemical reactions. During cutting of Ti-6Al-4V, adhesion occurring at the tool-chip interface is the main harbinger of the delamination problem.

At low cutting speeds, adhesive wear mode was observed during machining titanium alloys (Fig. 4.24). The location of the adhesion wear is greater in the case of Ti-6Al-4V compared to Ti-55531. For Ti-55531, the wear results show that the predominant degradation process is abrasion wear (Fig. 4.24b).

As shown by the EDS analysis illustrated in Fig. 4.28, particles debris are deposited on the tool surface during machining Ti-6Al-4V as successive layers when chips scroll to the surface. This leads to the adhesive wear mode. The latter

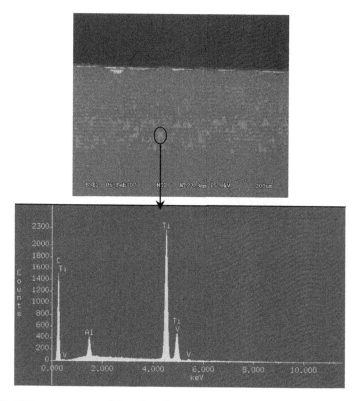

Fig. 4.28 EDS measurements of the adhered material (Ti-6Al-4V) on the rake face. V = 20 m/min, rake angle 0°

is highlighted by the change in the tool geometry and debonding of the coating (Fig. 4.25c).

Auger Electron Spectroscopy (AES) Surface Analysis was also used to verify the occurrence of diffusion process. This technique shows the evolution of the chemical composition at the interface between two different materials. Diffusion profiles were obtained along two lines located in the adhesion zone and inside the worn tool. The AES Surface Analysis performed on the cutting tool (Fig. 4.29a) show along L1 and L2 in Fig. 4.29b, c respectively the evolution of chemical species from the interface between the adhered material on the tool surface till some micrometers inside the tool substrate. It can be clearly showed from this figures that diffusion process occurred between the machined titanium alloys and cutting tool during machining. Diffusion profiles clearly show the diffusion of chemical species from the machined materiel (Ti, Al, V) to the cutting tool (W, Co, C) and vice versa. This diffusion process can be considered as a process which can generate adhesion wear of the cutting tool and/or delamination of coating.

During machining Ti-55531 at 65 m/min, a process of cracking starts resulting in a collapse of the cutting edge (Figs. 4.24d and 4.25d). This is produced by the

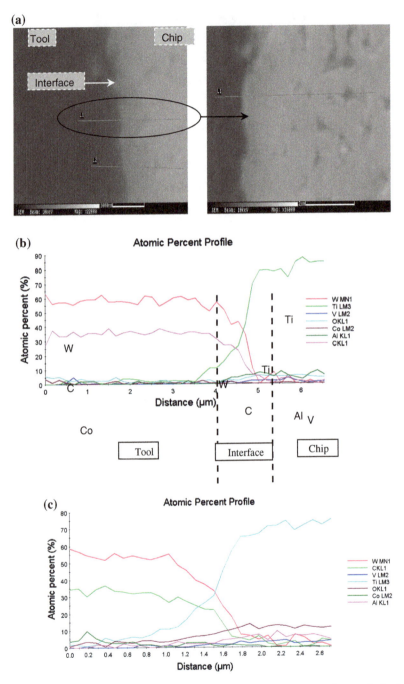

Fig. 4.29 Diffusion profiles obtained at the tool-chip interface by AES surface analysis along Lines L1 and L2 in the tool-chip interface and inside the cutting tool (Cutting conditions are identical to those of Fig. 4.28). **a** Localization of the analyzed worn zones inside the cutting tool (L1, L2). **b** AES surface analysis along line L1. **c** AES surface analysis along line L2

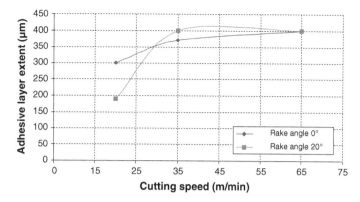

Fig. 4.30 Evolution of the adhesive layer extent on the tool surface during machining Ti-6Al-4V with different cutting speeds

combined effect of high pressures and large strain rates in the tool (substrate and coating). Adhesive wear layers on the tool surface were measured using a profilometer in the case of machining Ti-6Al-4V. Figure 4.30 shows the evolution of the adhesive layer extent versus cutting speed.

It appears from this result that the adhesive wear increases with the cutting speed and it stabilized at high speeds. This may due to the diffusion process between titanium alloy and the cutting tool at the interface.

In some cases and under the combined effect of pressure and temperature, welds are formed between tool and chip. With permanent mechanical stress (during machining), these welds are broken thereby causing chipping of the cutting surface. The irregularity on the tool surface can constitute attachment points for chip debris. By accumulating, they eventually form a detrimental macroscopic deposit on the cutting edge. This phenomenon has been observed for cutting tools when machining the Ti-6Al-4V alloy under low cutting speeds (see Figs. 4.24a and 4.25a).

Topographical survey using the profilometer (Fig. 4.31) confirmed that the tool wear exhibited when the machining of the Ti-55531 alloy changes the geometry of the cutting edge. Cracking is due to a fatigue phenomenon of the cutting edge followed by a break of the tool under cyclic loading during machining. The tool with 20° rake angle showed resistance to wear by fatigue and plastic deformation in the case of Ti-55531. Indeed, this geometry improves the flow of chips and material deformation.

In the case of Ti-6Al-4V, the average grain size is about 10 μm and the measured micro-hardness is in the range of 317 $HV_{0.2}$. For the Ti-55531, the microstructure is homogeneous and the grain size is about 1 μm with a microhardness of 379 $HV_{0.2}$. From a metallurgical point of view, the major difference between the Ti-6Al-4V and Ti-55531 is due to fineness of the microstructure. This generally leads to a higher mechanical strength. However this will strongly influence the machinability of the machined materiel and then tool wear as confirmed by Powell and Duggan [29] and by Arrazola et al. in [30]. In addition, the

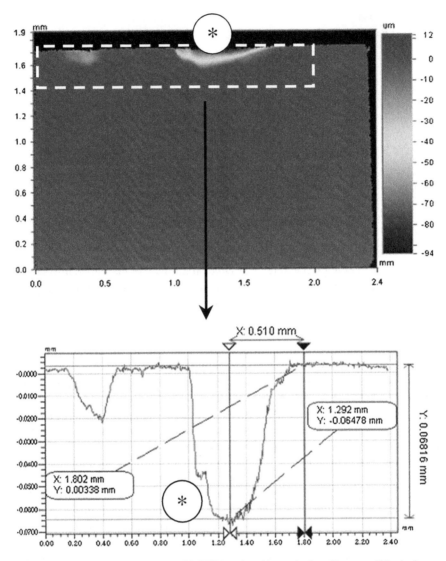

Fig. 4.31 Worn tool when machining Ti-55531 analyzed by optical profilometer. *Notched wear with depth of 61 μm

Ti-55531 is an alloy of β-type (near-β titanium alloy), with the presence of betagenic elements, such as chromium, for example, which limits the ability of the material to deform during machining.

Based on the results from the experimental tests we can give some ideas to improve the machining of the two alloys studied in this article. It is possible to increase the cutting speed in the case of machining Ti-6Al-4V, which causes the increase in temperature at the interface tool/material and therefore a significant softening of the machined material. Consequently cutting efforts will be lower. This solution can not be applied in the case of Ti-55531 because of its high strain rate sensitivity.

On the other hand, the improvement of the machining of Ti-6Al-4V can be carried out by changing the geometry of the tool. Increasing rake angle facilitates chip flow; this will cause reduction in cutting efforts and pressure. This improvement can also be applied to the machining Ti-55531 alloy.

4.4 Conclusion

In this research work, a detailed experimental approach for the comprehension of machining titanium alloys has been presented. Several analyzes (SEM, EDS, AES, Infrared camera, high-speed camera CCD and profilometer analysis) confirmed that differences in machinability and microstructure of tested titanium alloys can have an important impact on tool wear. The low machinability of titanium alloys due to the low thermal conductivity and high microhardness of these materials leads to severe and premature tool wear. The machining of titanium alloy Ti-55531 has been confronted with that of the Ti-6Al-4V alloy. To do this study, the mechanical, thermal, metallurgical and physico-chemical aspects have been deeply analyzed.

During the evolution of the machining efforts depending on cutting conditions and tool geometry, a decrease of cutting forces is noted for the case of Ti-6Al-4V when the cutting speed increases. This decrease is due to the thermal softening of the material under the effect of plastic deformation and cutting temperature. In other words, during the machining of Ti-6Al-4V, its yield stress decreases and therefore improves the machinability for this material. For the Ti-55531, this reduction has not been clearly identified. Unlike the Ti-6Al-4V alloys, the chip morphology of Ti-55531 shows a work hardening behavior. This means that a flow stress increases during machining making more difficult its machinability.

From a metallurgical point of view, the major difference between the Ti-6Al-4V and Ti-55531 is the fineness of the microstructure. The latter is automatically accompanied by a higher strength, better ductility and toughness. It can therefore be said that the Ti-55531 has a better fatigue resistance which greatly penalizes its machinability. Another difference from the Ti-55531 relative to the Ti-6Al-4V is the fact that the microstructure of Ti-55531 contains a volume fraction of the larger β phase. This increases its hardness while also reducing its machinability.

Results have also shown that the friction between the tool and the workpiece is close to 1 for the case of Ti-55531 alloy compared to Ti-6Al-4V. This means that during Ti-55531 machining the contact at the chip-tool interface is of a sticky type, thereby giving trend to the Ti-55531 alloy to more easily adhere to the cutting tool face. This type of contact also influences the thermal field which will be greater in the case of Ti-55531 (temperature for Ti-55531 about 800 °C, for Ti-6Al-4V about 670 °C). In addition, at the high temperatures, Ti-55531 alloy conducts heat better than the Ti-6Al-4V alloy. The low thermal conductivity of the Ti-6Al-4V promotes chip segmentation and therefore the reduction of machining efforts. For the Ti-55531 alloy, its higher conductivity allows it to further evacuate the stored heat, but at the same time it decreases its ability to soften and thereby deteriorates its machinability. Thus, it is found that the area affected by the wear in the case of Ti-55531 is greater than in the case of Ti-6Al-4V under the same machining conditions and for the same tool geometry.

The rest of our study also showed that the tool geometry has a significant influence on the machinability of both titanium alloys. With a rake angle of 0°, the difference of machinability is only about 12 % between Ti-55531 and Ti-6Al-4V and remains constant regardless of the cutting speed. With geometry of 20°, this difference is greater and reached high levels for high cutting speeds (approximately 34.7–65 m/min).

Regarding the coating, a difference of 8 % (20 m/min cutting speed) was found between the cutting forces when machining Ti-55531 alloy under the same conditions with coated and uncoated tools. This difference further increase as the cutting speed increases, it reaches 24 % for 65 m/min. The coating can therefore significantly improve the machinability of Ti-55531. However, further investigations with different types of coatings are necessary to understand the coating behavior for the optimization of titanium alloys machinability, specifically for the Ti-55531 alloy. The lubrication effect is more marked on the tool damage (with or without coating). Thus, the analysis showed that the lubrication indirectly affects the machinability of titanium alloys through tool wear. This finding differs from the effect of the coating which shows a direct effect on machinability and indirect effect on tool wear.

References

1. Boyer RR (1996) An overview on the use of titanium in the aerospace industry. Mater Sci Eng 213A:103–114
2. Ginting A, Nouari M (2009) Surface integrity of dry machined titanium alloys. Int J Mach Tools Manuf 49(3–4):325–332
3. Clément N, Lenain A, Jacques PJ (2007) Mechanical property optimization via microstructural control of new metastable beta Titanium alloys, processing and characterizing Titanium alloys overview. JOM 59:50–53
4. Clement N et al (2005) In: JM Howe et al (ed) Proceedings of international conference solid-solid phase transformations in inorganic materials. TMS, Warrendale, PA, pp 603–608
5. Nouari M, Ginting A (2006) Wear characteristics and performance of multi-layer CVD-coated alloyed carbide tool in dry end milling of titanium alloy. Surf Coat Technol 200(18–19):5663–5676
6. Ginting A, Nouari M (2006) Experimental and numerical studies on the performance of alloyed carbide tool in dry milling of aerospace material. Int J Mach Tools Manuf 46(7–8):758–768
7. Nouari M, Makich H (2013) Experimental investigation on the effect of the material microstructure on tool wear when machining hard titanium alloys: Ti-6Al-4V and Ti-555. Int J Refract Metal Hard Mater 41:259–269
8. Ginting A, Nouari M (2007) Optimal cutting conditions when dry end milling the aeroengine material Ti-6242S. J Mater Process Technol 184:319–324

9. Komanduri R (1981) Turkovich B.F.V., new observations on the mechanism of chip formation when machining titanium alloys. Wear 69:179–188
10. Ezugwu EO, Wang ZM (1997) Titanium alloys and their machinability—a review. J Mater Process Technol 68:262–274
11. Subramanian SV, Ingle SS, Kay DAR (1993) Design of coatings to minimize tool crater wear. Surf Coat Tech 61:293–299
12. Bouchnak TB (2010) Etude du comportement en sollicitations extrêmes et de l'usinabilité d'un nouvel alliage de titane aéronautique, PhD thesis, Ref. 2010-ENAM-0051, Arts et Métiers ParisTech—Centre d'Angers
13. Fanning JC (2005) Properties of TIMETAL 555 (Ti-5Al-5Mo-5 V-3Cr-0.6Fe). JMEPEG 14:788–791
14. Nyakana SL, Fanning JC, Boyer RR (2005) JMEPEG 14:799–811
15. Semiatin SL (1999) Seetharaman V, Ghosh AK (1999) Plastic flow, microstructure evolution, and defect formation during primary hot working of titanium and titanium aluminide alloys with lamellar colony microstructures. Philos Trans R Soc A: Mathe Phys Eng Sci 357(1756): 1487–1512
16. Jackson M, Dashwood R, Christodoulou L, Flower H (2005) The microstructural evolution of near beta alloy Ti-10 V-2Fe-3Al during subtransus forging. Metall Mater Trans A 36:1317–1327
17. Benedetti M, Fontanari V (2004) The effect of bi-modal and lamellar microstructures of Ti-6Al-4V on the behaviour of fatigue cracks emanating from edge-notches. Fatigue Fract Eng Mater Struct 27:1073–1089
18. Nouari M, Calamaz M, Girot F (2008) Mécanismes d'usure des outils coupants en usinage à sec de l'alliage de titane aéronautique Ti-6Al-4V, C.R. Mécanique 336:772–781
19. Devillez A, Schneider F, Dominiak S, Dudzinski D, Larrouquere D (2007) Cutting forces and wear in dry machining of Inconel 718 with coated carbide tools. Wear 262(7–8):931–942
20. Singh Gill S, Singh R, Singh H, Singh J (2011) Investigation onwear behaviour of cryogenically treated TiAlN coated tungsten carbide inserts in turning. Int J Mach Tools Manuf 51(1):25–33
21. Castanho J, Vieira M (2003) Effect of ductile layers in mechanical behaviour of TiAlN thin coatings. J Mater Process Technol 143:352–357
22. Battagliaa JL, Coisb O, Puigsegura L, Oustaloupb A (2001) Solving an inverse heat conduction problem using a non-integer identified model. Int J Heat Mass Transfer 44:2671–2680
23. Puerta Velasquez JD, Bolle B, Chevrier P, Geandier G, Tidu A (2007) Metallurgical study on chips obtained by high speed machining of a Ti–6 wt%Al–4 wt%V alloy. Mater Sci Eng A 452–453, 469–474
24. He Yi (2005) Rapid thermal conductivity measurement with a hot disk sensor: part 1. Theoret Considerations Thermochim Acta 436:122–129
25. Abdel-Aal HA, Nouari M, Mansori ELM (2009) Tribo-energetic correlation of tool thermal properties to wear of WC-Co inserts in high speed dry machining of aeronautical grade titanium alloys. Wear 266:432–443
26. Merchant E (1945) Mechanics of the metal cutting process II. Plasticity conditions in orthogonal cutting. J Appl Phys 16:318–324
27. Merchant E (1945) Mechanics of the metal cutting process I. Orthogonal cutting and a type 2 chip. J Appl Phys 16:267–275
28. Komanduri R (1982) Some clarifications on the mechanics of chip formation when machining titanium alloys. Wear 76:15–34
29. Powell BE, Duggan TV (1986) Predicting the onset of high cycle fatigue damage: an engineering application for long crack fatigue threshold data. Int J Fatigue 8:187–194
30. Arrazola P-J, Garay A, Iriarte L-M, Armendia M, Marya S, Le Maître F (2009) Machinability of titanium alloys (Ti6Al4 V and Ti555.3). J Mater Process Technol 209:2223–2230

Chapter 5
Green Machining of Ti-6Al-4V Under Minimum Quantity Lubrication (MQL) Condition

Liu Zhiqiang

Abstract Although many studies have been done on the MQL applications in different machining process, there are few of studies on the influences about selections of MQL parameters. Different parameters of MQL system have different influences on the milling force and milling temperature, which are closely connected to the lubrication and the coolant. The cutting force and temperature play significant roles in the improving/reducing cutting quality of workpiece and extend/shorten tool life, by changing different parameters of MQL system. This chapter presents an experiment of end-milling titanium alloy with MQL system, discussing the influences of different parameters. One object of the experiment is to investigate the influences of Ti-6Al-4V. The results of experiment will help to understand the influences of selecting different parameters on the end-milling process. Another object is to apply the selected optimal parameters in the former object to milling Ti-6Al-4V.

5.1 Introduction

From an aim in the reduction of environment and economic cost, there are critical needs to reduce the use of cutting fluid in cutting process [1–5]. Minimum quantity lubrication (MQL) is to supply a minute quantity of cooling lubricant medium (air–oil mist) to the cutting zone. The mist (air–oil mist) sprayed by MQL system plays a very important role, which is used to cool the tool, workpiece and machine tool; and acts as a lubricant at the interface of the tool and the chip. The air–oil mixture fed onto the cutting zone also called aerosol or oil-fog. The oil applied in the MQL device for machining process is biodegradable vegetable oil derivatives [3, 6, 7], which has extensive friction-reducing properties. The particle sizes of oil-

L. Zhiqiang (✉)
School of Mechanical Engineering, Jiangsu University of Science and Technology,
Zhenjiang 212003, People's Republic of China
e-mail: Liuzhiqiang09@gmail.com

J. P. Davim (ed.), *Machining of Titanium Alloys*,
Materials Forming, Machining and Tribology, DOI: 10.1007/978-3-662-43902-9_5,
© Springer-Verlag Berlin Heidelberg 2014

mist, generated by MQL device, range from sub-micrometer to over 50 μm in diameter [8, 9].

Most of literatures [3, 4, 6, 10–14] had great effort to compare dry cutting, wed cutting and cutting with MQL. Many literatures had studied the tool-wear and tool-life in different machining process' including turning [15], milling [11, 16], drilling [13, 17], grinding [2, 18, 19], grooving [20], reaming [6], etc. Also many cutting process, applied with MQL system, of different materials were studied [1–3, 14–17, 19–22], including aluminum [13], titanium, inconel [15, 22] or super-alloy, steel [1, 10, 21, 23], etc.

However, few of literatures [11, 18] paid attention to how to select the MQL parameters. Literature [11] studied the influences on milling with different nozzle locations and directions by applying the CFD method. Literature [18] studied the influence of mist parameters on MQL grinding process. This chapter focus on analyzing influences on the cutting force and temperature by changing different parameters of MQL system. The various parameters of minimum quantity lubrication (MQL), including air pressure, oil quantity, nozzle location and direction, have great influences on the cutting force and the cutting temperature. Especially, the method of selecting parameters acts as a significant role on the different processes of metal cutting.

5.2 Milling Force and Temperature Model

In order to apply the MQL system effectively, the cutting, lubrication and cool mechanisms in the condition of MQL need to be clearly understood. Different parameters of MQL system have different influences on the cutting force and cutting temperature, which are closely connected to lubrication, thermal generation and thermal dissipation.

5.2.1 Dynamic Force in End Milling

5.2.1.1 Friction Force

The milling resultant force on the cutting insert can be decomposed into two components: friction F_f and normal pressure F_n, which are acting on the rake face of milling cutter. Friction F_f, is closely connected to cutter's wear and cutting temperature; and normal force F_n also plays an important role in milling cutter's wear.

Under Minimum Quantity Lubrication condition, there is a type of boundary lubrication. The oil consists of long-chain polar molecules that are physically absorbed on the high points of mating surfaces. The oil lubrication film is not complete and the contact area is changing between tool-workpiece and chip-tool in the milling process.

Fig. 5.1 Sliding friction
region and sticking friction
region

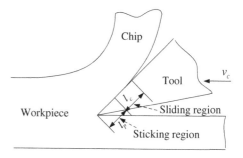

In the contact zone of tool-chip shown in Fig. 5.1, there are two regions of friction: sliding friction region $F_{sliding}$, and sticking (adhesive) friction region $F_{adhesive}$ [24]. The friction force contains three parts following as:

$$F_f = F_{adhesive} + F_{sliding} + F_{plowing} \tag{5.1}$$

where, $F_{adhesive}$ is the adhesive internal-friction force in the sticking region, $F_{sliding}$ is the sliding friction force in the sliding region, $F_{plowing}$ is the plowing force due to deformation, which could be ignored due to the smaller value.

In the sliding friction, the friction force can be expressed as followed:

$$F_{sliding} = \mu_f F_n = K_f \cdot A_{sliding} \tag{5.2}$$

where, μ_f is the average friction coefficient on the rake face ($\mu_f = \tan \beta$, β is the friction angle). K_f is the instantaneous cutting pressure coefficients for the friction component. $A_{sliding}$ is the area of sliding region.

In the sticking friction, the friction force can be expressed as followed:

$$F_{adhesive} = \tau \cdot A_{adhesive} \tag{5.3}$$

where, τ is the shear strength of the internal-friction material. $A_{adhensive}$ is the area of sticking region.

Substituting Eqs. (5.2) and (5.3) into Eq. (5.1), the friction force can be obtained:

$$F_f = \tau \cdot A_{adhesive} + K_f \cdot A_{sliding} \tag{5.4}$$

As shown in Fig. 5.1, the length of the sticking region l_p in the two-dimensional is assumed to be equal to the length of the sliding region l_c [25, 26]. Thus, in the end-milling process using small helix angle inserts, $A_{adhesive} \approx A_{sliding} \approx \frac{1}{2}A_{contact}$. The friction can be calculated as

$$F_f = \frac{1}{2}A_{contact}(\tau + K_f) = K_{fe} \cdot A_{contact} \tag{5.5}$$

Fig. 5.2 Equivalent cutting
edge model

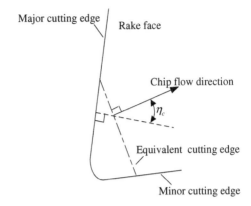

where, $A_{contact}$ is the contact area of the too-chip. $K_{fe} = \frac{1}{2}(\tau + K_f)$ is the equivalent
pressure coefficient for the friction component.

5.2.1.2 Equivalent Cutting Edge Model

Based on the studies [27] of Colwell, Stabler, Armarego, etc., Hu et al. [28]
presented a geometric model called equivalent cutting edge model. Considering
the influence of major cutting edge and minor cutting edge, the chip flow direction
is along the direction shown in the Fig. 5.2.

The friction can be assumed along the chip flow direction on the rake face
approximately. Therefore, on the race face, the friction can be decomposed into
two components as followed:

$$\begin{pmatrix} F_{f1} \\ F_{f2} \end{pmatrix} = \begin{pmatrix} F_f \cos \eta_c \\ F_f \sin \eta_c \end{pmatrix} \qquad (5.6)$$

5.2.1.3 Transformation Between Measuring Force and Dynamic Milling Force

According the above analysis of friction, the cutting force can be given as
followed.

$$F_f = K_{fe} \cdot A_c \qquad (5.7)$$

$$F_n = K_n \cdot A_c \qquad (5.8)$$

where, $K_{fe} = \frac{1}{2}(\tau + K_f)$ is the equivalent instantaneous cutting pressure coefficient
for the friction component. K_n is the instantaneous cutting pressure coefficients for
the normal component. Meanwhile, in the end milling process with various MQL
parameters, K_{fe} and K_n could be changed relevantly.

Fig. 5.3 Angular position θ shown in milling geometry

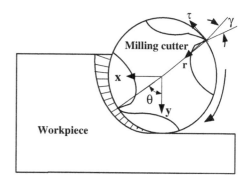

The chip cross sectional area A_c is given as followed [29]:

$$A_c = \left(f_z \sin \theta + R - \sqrt{R^2 - (f_z \cos \theta)^2} \right) \cdot a_p \qquad (5.9)$$

where, a_p is the depth of cut, and f_z, R are the feed per tooth per revolution and milling cutter radius, respectively, and θ is the instantaneous angular position along with the cutter's rotation, as shown in Fig. 5.3. Thus, $f_z \sin \theta + R - \sqrt{R^2 - (f_z \cos \theta)^2}$ is the instantaneous uncut chip thickness.

The milling force is also decomposed into three components F_x, F_y and F_z measured in the global Cartesian coordinate system. A transform matrix $\mathbf{T}$ is used to transform F_f and F_n into global Cartesian coordinate system. The transform equation is given by,

$$\begin{pmatrix} F_x \\ F_y \\ F_z \end{pmatrix} = \mathbf{T} \cdot \begin{pmatrix} F_n \\ F_f \cos \eta_c \\ F_f \sin \eta_c \end{pmatrix}, \quad \text{and} \quad \begin{pmatrix} F_n \\ F_f \cos \eta_c \\ F_f \sin \eta_c \end{pmatrix} = \mathbf{T}^{-1} \cdot \begin{pmatrix} F_x \\ F_y \\ F_z \end{pmatrix} \qquad (5.10)$$

where, η_c is the chip flow angle with respect to the radial force in the axial rake plane. Meanwhile, milling forces are different at each time step for the same mill and workpiece in the same milling process. Therefore, the measuring forces in the experiment are discredited according to the time step.

The milling force is a dynamic force due to the transient variation cross-selection area of uncut chip. Dynamic milling force model were studied by many researchers [29–31], such as Attintas, Shin, Davies.

Therefore, the transform matrix is given by,

$$\mathbf{T} = \begin{pmatrix} -\cos \theta & \sin \theta & 0 \\ -\sin \theta & -\cos \theta & 0 \\ 0 & 0 & 1 \end{pmatrix} \begin{pmatrix} \cos \gamma & \sin \gamma & 0 \\ -\sin \gamma & \cos \gamma & 0 \\ 0 & 0 & 1 \end{pmatrix} \begin{pmatrix} \cos \varphi & 0 & \sin \varphi \\ 0 & 1 & 0 \\ -\sin \varphi & 0 & \cos \varphi \end{pmatrix}$$

$$(5.11)$$

where, γ is the radial rake angle, and φ is helix angle of the end mill.

Thus, K_{fe}, K_n and η_c are obtained from the measured forces F_x, F_y and F_z by the following transformation:

$$\begin{pmatrix} F_n \\ F_f \cos\eta_c \\ F_f \sin\eta_c \end{pmatrix} = \begin{pmatrix} K_n \\ K_{fe}\cos\eta_c \\ K_{fe}\sin\eta_c \end{pmatrix} \cdot A_c = \mathbf{T}^{-1} \cdot \begin{pmatrix} F_x \\ F_y \\ F_z \end{pmatrix} \qquad (5.12)$$

With the various MQL parameters, K_{fe} and K_n could be changed relevantly.

In the milling process, various MQL parameters have special effects on F_f and F_n, which are instantaneous varying values. According to the Oxley equation (1967) [32], shear angle ϕ will increase with the friction angle β decrease, vice versa.

5.2.2 Milling Heat and Temperature

In the milling process, the friction and the deformation change with different MQL parameters. Actually, the shear angle ϕ, the friction angle β and adhesive friction change along with different MQL parameters due to the different effects of lubrication and heat dissipation. Meanwhile, the wear of milling cutter is reduced by the influence of micro-lubrication. Thus, the heat is also reduced due to the reducing tool's wear.

5.2.2.1 Heat Generation

During the cutting process, there are three heat sources [24], including primary heat source Q_1 produced by plastic shear deformation in the primary shear zone, secondary heat source Q_2 produced by plastic deformation due to shearing and friction on the cutting face, third heat source Q_3 produced by friction between workpiece and tool on the tool flank. Three heat sources are shown in Fig. 5.4.

Moreover, cutting lubrication oil, used in the MQL, has good lubrication performances. Thus, the heat produced by friction will be reduced obviously using the MQL system. Generally, cutting lubrication oil is un-saturation with a long-chain molecule structure. Unlike mineral-based and synthetic fluids, the oil of MQL is extracted from vegetable-based fluids are developed in the presence of oxygen, which causes them to be bond with the surface of metals better [3, 7]. Therefore, it will provide superior lubricity and the friction between the contact zones of chip-tool and tool-workpiece will be reduced.

5.2.2.2 Heat Dissipation

In general, the oil fog of MQL system has a significant role in lubrication and heat dissipation. Compressed air with oil fog is sprayed with lower temperature due to

Fig. 5.4 Heat sources

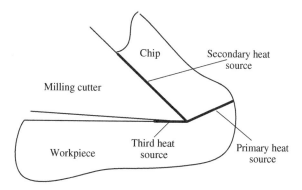

the work done by air-inflating. During this process, the workpiece and tool are cooled due to the enhanced convection of heat transfer.

On the other hand, the oil fog, which permeates and enters the zones of tool-chip and tool-workpiece, is boiled and gasified to absorb heat. During the cutting process, the heat transferred to the droplets of oil-fog raise the temperature above its boiling point and vaporizes the fluid [9, 12].

The dynamic viscosity μ of gas–liquid (oil–mist) two phases is given by:

$$\mu = \mu_{liq} - (\mu_{liq} - \mu_{gas}) \frac{1}{r + 1} \qquad (5.13)$$

where, μ_{liq} is the dynamic viscosity of liquid (oil), μ_{gas} is dynamic viscosity of gas (air), and $r = Q_{air}/Q_{oil}$ is the gas liquid ratio (the ratio of air quantity to oil quantity).

According to the dynamic viscosity's formula, which is closed with interior-frictional fluid property, the friction will changed along with the value of dynamic viscosity μ. The cooling effect of the fluid with lower dynamic viscosity is better than the one with higher dynamic viscosity. The fluid with higher dynamic viscosity has bigger thermal resistance. Seen from the Eq. (5.13), $\mu < \mu_{liq}$, its effect of temperature reduction is better than the single-phase of oil.

5.3 Comparison of Effects with Different MQL System Parameters

5.3.1 Milling Experimental Procedures Using MQL System

5.3.1.1 Workpiece Material

The workpiece material was titanium alloy Ti-6Al-4V tested in rectangular blocks of $30 \times 20 \times 10$ mm^3. Ti-6Al-4V is a kind of widely used titanium alloy in

Table 5.1 Chemical composition of Ti-6Al-4V

Composition	Ti	Al	V	Fe	C	N	H	O
Containing (in wt%)	The balance	5.5 ~ 6.8	3.5 ~ 4.5	0.30	0.10	0.05	0.015	0.20

Table 5.2 Mechanical properties of Ti-6Al-4V

Titanium alloy	Ultimate tensile strength (MPa)	Elongation (%)	Hardness (HB)	Tensile strength 400 °C (MPa)
Ti-6Al-4 V	1,000	14	241	550

industry. The Chemical composition and the mechanical properties of Ti-6Al-4V are displayed in Tables 5.1 and 5.2.

5.3.1.2 Milling Experiment and MQL Supply Device

Milling tests were performed using the machining center MAHO 600C. The cutting tool used for the milling was an index-able end milling cutter (Mill 1-18, EDT1805) with 25 mm diameter manufactured by Kennametal Inc. (USA). The end mill was equipped with only one insert for convenience of cutting modeling analysis. The insert with coating (PVD TiAlN) was applied in the experiments.

The milling machine tool was equipped with an external MQL system (BLU-EBE FK, FUJI BC ENGINEERING, JAPAN). The principle of external MQL system is shown in Fig. 5.5. The oil (LB-1) is a kind of oil based fluid extracted from vegetable oil.

As shown in Fig. 5.6, the cutting forces were measured by Kistler dynamometer (model 9272) with the sampling frequency 1,000 Hz. Kistler 5019B four-channel charge amplifier was used to enlarge the signal and the corresponding software DynoWare was used to save and analyze the data. The Cutting Temperature Measurement method Based on Semi-manual Thermocouple Method.

5.3.1.3 Milling Test Procedures

In order to acquire the optimal spray parameters of MQL system, single-factor experiments of MQL system were done, in which only one factor was changed every time. Milling force and Milling temperature were measured in the experiments. The location of spray nozzle have great influences on the milling process. The location of spray nozzle is positioned with two factors, including spray target distance with the milling tool (D_0, mm) and spray angle to feed direction (θ, degree). These two factors are also shown in Fig. 5.7.

At first, pressure of air P_0 was changed every time. Then, spray distance D_0 was changed. And spray angle to feed direction θ with respect to tool feed direction

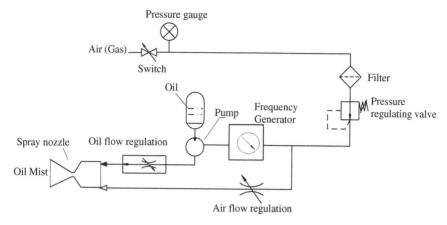

Fig. 5.5 The principle of external MQL system

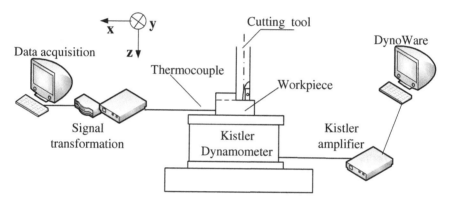

Fig. 5.6 Sketch of experimental setup

Fig. 5.7 The location of
spray nozzle

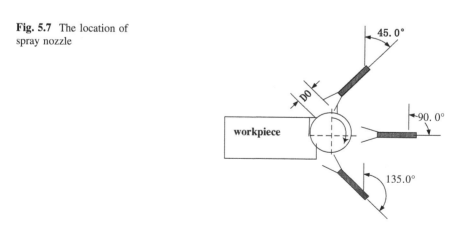

Table 5.3 Single-factor experiments of MQL system

No./ factor	Air pressure P_0 (Mpa)	No./ factor	Spray distance D_0 (mm)	No./ factor	Spray angle to feed direction θ (degree)	No./ factor	Oil delivery Q_{oil} (ml/h)
1	0.1	1	15	1	45	1	2
2	0.3	2	25	2	90	2	4
3	0.5	3	35	3	135	3	6
4	0.7	4	45			4	8
						5	10

was followed to be changed. Finally, oil delivery Q_{oil} was changed, as shown in Table 5.3. Meanwhile, the air for the MQL application was applied at the constant rate of 125 l/min (l/min).

With the different air-pressures of MQL system shown in Table 5.3 while the other parameters remain unchanged. Similarly, the other factors including spray distance, spray angle to feed direction, Oil delivery were changed, as shown in Table 5.3.

In order to avoid additional influence, the cut parameters of milling process were remain unchanged in each single-factor experiments of MQL system. The milling tests were performed with the cutting speed 60 m/min and a feed rate of 0.05 mm per tooth. During each milling cycle, depth of cut and width of cut were 5 and 1 mm, respectively.

5.3.2 Effects of MQL System Parameters

5.3.2.1 Influence of Air-Pressure

With the different air-pressures of MQL system shown in Table 5.3 while the other parameters are constant (Spray distance $D_0 = 20$ mm, Spray angle to feed direction $\theta = 45°$, Oil delivery $Q_{oil} = 4$ ml/h), the different milling forces and milling temperatures are obtained. Firstly, the instantaneous cutting pressure coefficients for friction component K_{fe} and for normal component K_n are calculated from the measured data according to the Eq. (5.12), as shown in Figs. 5.8 and 5.9. Under the conditions with the same MQL parameters, the two cutting force coefficients (tangential direction K_{fe} and Normal direction K_n) will reduced along with the increasing of chip thickness, which show the size effects of extrusion and plowing became clear, cutting size effect increased.

As seen from Figs. 5.8 and 5.9, two coefficients (K_{fe} and K_n) are changing along with different air-pressures. According to Eqs. (5.7) and (5.8), these two coefficients have closely connection with the friction load and normal load above the rake face of the milling cutter.

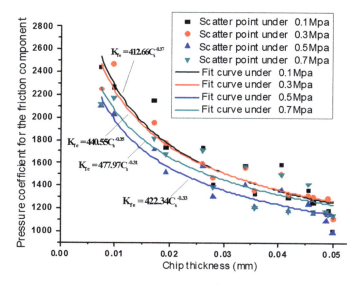

Fig. 5.8 The instantaneous coefficient (K_{fe}) with different air-pressures

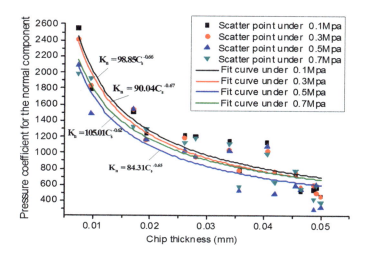

Fig. 5.9 The instantaneous coefficient (K_n) with different air-pressures

Especially, two coefficients K_{fe} and K_n with the air-pressure 0.5 Mpa are less than other values in other air-pressures' condition. The reason is that the air-pressure has the important role in the changing diameters of oil-mist. The diameters of oil-mist with the lower air pressures are bigger than ones with high air-pressure.

The droplet's diameter of air-mist is determined by the fluid flow speed [8]. The diameter of droplets is reduced along with the increasing pressure.The droplets with small diameter have better ability to penetrate into the contact zones of chip-tool and tool-workpiece.

On the other hand, the too high air-pressure has adverse effect on the oil-mist. According to the conservation of momentum and mechanical Energy, too high velocity will apt to rebounding. Because higher air-pressure produces the droplets with the higher velocity, it will lead to the rebounding of droplets with the workpiece and cutter; and the lower effect on the surface of metal to form oil film by physical adsorption.

It is known that friction and normal force lead to different wear of cutter. Thus, the air-pressure should be adjusted in the cutting process with different cutting parameters, different materials of workpieces or different cutters.

5.3.2.2 Influence of Spray-Distance

Similar to the ways of adjusted air-pressure, the spray-distance of nozzle was changed with different values in order to analyze the influence on the cutting force and temperature. With the different spray-distances of nozzle shown in Table 5.3 while the other parameters remain unchanged, the different milling forces and milling temperatures are obtained. Firstly, the instantaneous cutting pressure coefficients for friction component K_{fe} and for normal component K_n are calculated from the measured data according to the Eq. (5.12), as shown in Figs. 5.10 and 5.11.

Similarly, as shown in Figs. 5.10 and 5.11, two coefficients (K_{fe} and K_n) are changing along with different air-pressures. Especially, two coefficients K_{fe} and K_n with the spray-distance 25 mm are less than other values in other spray-distances' condition. The reason is that the spray-distance also has an great influence on the velocity and diameters of the oil-mist droplets. The reason is the same as the air-pressure. Near distance is benefit for immersion and penetrate of the oil-mist droplets.

According to the famous Bernoulli's equation, the fluid flow speed is raised along with the decreasing hydraulic head. The hydraulic head equals to the spray-distance. Thus, the diameter of droplets is reduced along with the increasing fluid flow speed. In the same way, the droplets with small diameter have better ability to penetrate into the contact zones of chip-tool and tool-workpiece. However, because of the rebounding of the droplets encountering obstacles, too near distance also has adverse effect on the oil-mist. Likewise, it will lead to the lower effect of physical adsorption of oil film.

5.3.2.3 Influence of Oil-Quantity

The gas liquid ratio (the ratio of air quantity with oil quantity) is another important factor affecting the velocity and the diameters of oil-mist droplets. The value of

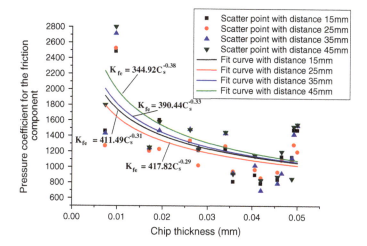

Fig. 5.10 The instantaneous coefficient (K_{fe}) with different spray-distances

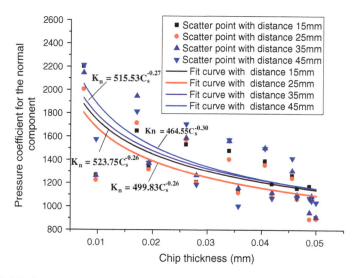

Fig. 5.11 The instantaneous coefficient (K_n) with different spray-distances

gas liquid ratio should be limited in an appropriate range in order to get good effects.

With the increase in oil-quantity of MQL, cutting temperature and cutting force are decreased in a slow way, as shown in Fig. 5.12. The reason is that the number of oil-particles entering the contact zone of tool-chip is increasing. Thus, the lubrication will be improved as well as friction reduction. Meanwhile, the increasing of oil-quantity is good for the cooling.

Fig. 5.12 Cutting forces and cutting temperature with different oil-quantities

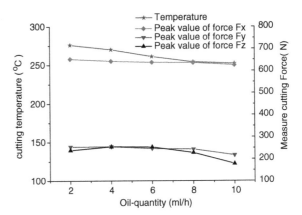

Fig. 5.13 Cutting force and temperature with different relative positions of nozzle

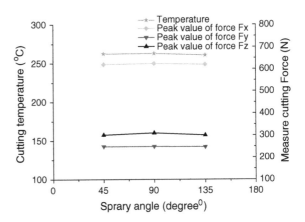

But, when the amount of oil-quantity reaches to 10 ml/h approximately, the reduction of cutting temperature and the cutting force is no longer obvious. One reason is that the value of gas liquid ratio is limited in a range which is benefit for getting appropriate velocity and diameters of oil-mist droplets. Another reason is that the boundary lubricating film at this gas liquid ratio has been stable, and then increased of the oil-quantity will not improve the lubrication.

5.3.2.4 Influence of Spray-Angle

Seen from Fig. 5.13, the cutting force and temperature show that the value with the spray-angle 135° is a little lower than the values with other two angles. The appearance of different spray angles is also studied in the literature [11]. These differences are due to the reason that in the 45° and 90° position the oil-mist does not completely penetrate into the inner zones of the tool edges. And the oil-mist in

135° position penetrates into the inner zones of the tool edges working in a very efficient way [11].

5.4 Conclusions

This chapter introduces the influences of the MQL parameters on the cutting force and temperature, which is closely connected to the tool life and cutting quality of workpiece. The effect of the MQL in the milling of titanium has been presented, and three main conclusions can be drawn:

(1) Because of the influences on the diameters of oil-mist, the air-pressure and spry-distance should be adjusted in the cutting process with different cutting parameters, different materials of workpieces or different cutters. Meanwhile, too low or high air-pressure has not benefit for the penetrate ability of oil-mist.

(2) The spry-angle of nozzle position in relation to feed direction has minor effect on the cutting force and cutting temperature.

(3) With the increase of oil-quantity of MQL, cutting temperature and cutting force are decreased in a slow way. But, when the amount of oil-quantity reach to a value, the reduction of cutting temperature and the cutting force is no longer obvious. It helps to clearly understand that the reduction of the quantity of cutting oil in the machining process is possible, which leads to lower machining costs as well as a little decreasing of the cutting quality.

Acknowledgements The work is supported by the National Natural Science Foundation of China (No. 51305174).

References

1. Dhar NR, Kamruzzaman M, Ahmed M (2006) Effect of minimum quantity lubrication (MQL) on tool wear and surface roughness in turning AISI-4340 steel. J Mater Process Technol 28:299–304

2. Bin S, Albert JS, Simon CT (2008) Application of nanofluids in minimum quantity lubrication grinding. Tribol Trans 51:730–737

3. Weinert K, Inasaki I, Sutherland JW, Wakabayashi T (2004) Dry machining and minimum quantity lubrication. CIRP Ann Manuf Technol 53(2):11–537

4. Tawakoli T, Hadad MJ, Sadeghi MH, Daneshi A, Stöckert S, Rasifard A (2009) An experimental investigation of the effects of workpiece and grinding parameters on minimum quantity lubrication—MQL grinding. Int J Mach Tools Manuf 49(12–13):924–932

5. Davim PJ, Astakhov VP (2008) Machining fundamentals and recent advances [M]. Springer, London, pp 195–223

6. De Chiffre L, Tosello G, Píška M, Müller P (2009) Investigation on capability of the reaming process using minimal quantity lubrication. CIRP J Manuf Sci Technol 2(1):47–54

7. Boelkins C, Unist (2009) MQL: Lean and Green. Cutting Tool Engineering 61(3):71–75
8. Ju C (2005) Development of particulate imaging systems and their application in the study of cutting fluid mist formation and minimum quantity lubrication in machining. pp 154–156
9. Sun J (2004) Cutting fluid mist formation and behavior mechanisms. Michigan Technological University, Houghton
10. Liao YS, Lin HM, Chen YC (2007) Feasibility study of the minimum quantity lubrication in high-speed end milling of NAK80 hardened steel by coated carbide tool. Int J Mach Tools Manuf 47(11):1667–1676
11. López de Lacalle LN et al (2006) Experimental and numerical investigation of the effect of spray cutting fluids in high speed milling. J Mater Process Technol 172:11–15
12. Byrne G, Dornfeld D, Denkena B (2003) Advancing cutting technology. CIRP Ann Manuf Technol 52(2):483–507
13. Bhowmick S, Alpas AT (2008) Minimum quantity lubrication drilling of aluminium–silicon alloys in water using diamond-like carbon coated drills. Int J Mach Tools Manuf 48(12–13):1429–1443
14. Varadarajan AS, Philip PK, Ramamoorthy B (2002) Investigations on hard turning with minimal cutting fluid application (HTMF) and its comparison with dry and wet turning. Int J Mach Tools Manuf 422(2):193–200
15. Obikawa T, Asano Y, Kamata Y (2009) Computer fluid dynamics analysis for efficient spraying of oil mist in finish-turning of Inconel 718. Int J Mach Tools Manuf 49(12–13):971–978
16. Rahman M, Senthil Kumar A, Salam MU (2002) Experimental evaluation on the effect of minimal quantities of lubricant in milling. Int J Mach Tools Manuf 42(5):539–547
17. Bhowmick S, Lukitsch MJ, Alpas AT (2010) Dry and minimum quantity lubrication drilling of cast magnesium alloy (AM60). Int J Mach Tools Manuf (In Press, Corrected Proof, Available online 10 February 2010)
18. Tawakoli T, Hadad MJ, Sadeghi MH (2010) Influence of oil mist parameters on minimum quantity lubrication-MQL grinding process. Int J Mach Tools Manuf (In Press, Accepted Manuscript, Available online 15 March 2010)
19. da Silva LR, Bianchi EC, Fusse RY, Catai RE, Franca TV, Aguiar PR (2007) Analysis of surface integrity for minimum quantity lubricant-MQL in grinding. Int J Mach Tools Manuf 47(2):412–418
20. Obikawa T, Kamata Y, Shinozuka J (2006) High speed grooving with applying MQL. Int J Mach Tools Manuf 46(14):1854–1861
21. Dhar NR, Ahmed MT, Islam S (2007) An experimental investigation on effect of minimum quantity lubrication in machining AISI 1040 steel. Int J Mach Tools Manuf 5:748–753
22. Obikawa T, Kamata Y, Asano Y, Nakayama K, Otieno AW (2008) Micro-liter lubrication machining of Inconel 718. Int J Mach Tools Manuf 48(15):1605–1612
23. Thepsonthi T, Hamdi M, Mitsui K (2009) Investigation into minimal-cutting-fluid application in high-speed milling of hardened steel using carbide mills. Int J Mach Tools Manuf 49(2):156–162
24. Shaw MC (2005) Metal cutting principles [m]. Oxford University Press, New York, pp 156–158
25. Li X (1997) Development of a predictive model for stress distributions at the tool-chip interface in machining. J Mater Process Technol 63(1–3):169–174
26. Tay AO, Stevenson MG, de Vahl Davis G (1974) Using the finite element method to determine temperature distributions in orthogonal machining. Proc Inst Mech Eng 188(55):627–638
27. Adibi-Sedeh AH, Madhavan V, Bahr B (2002) Upper bound analysis of oblique cutting with nose radius tools. Int J Mach Tools Manuf 42(9):1081–1094

28. Hu RS, Mathew P, Oxley PLB, Young HT (1986) Allowing for end cutting edge effects in predicting forces in bar turning with oblique machining conditions. Proceedings of the Institution of Mechanical Engineers. Part C: J Mech Eng Sci 200(C2):89–99
29. Li H, Shin YC (2006) A comprehensive dynamic end milling simulation model. J Manuf Sci Eng 128(1):86–94
30. Rao BC (2002) Modeling and analysis of high speed machining of aerospace alloys. Purdue University, West Lafayette, pp 25–66
31. Altintas Y, Eynian M, Onozuka H (2008) Identification of dynamic cutting force coefficients and chatter stability with process damping. CIRP Ann Manuf Technol 57(1):371–374
32. Oxley PLB (1989) Mechanics of machining: an analytical approach to assessing machinability. Wiley, New York

Chapter 6
Ultrasonically Assisted Machining of Titanium Alloys

Anish Roy and Vadim V. Silberschmidt

Abstract In this chapter we discuss the nuances of a non-conventional machining technique known as ultrasonically assisted machining, which has been used to demonstrate tractable benefits in the machining of titanium alloys. We also demonstrate how further improvements may be achieved by combining this machining technique with the well known advantages of hot machining in metals and alloys.

6.1 Introduction

Recently, machinability of titanium alloys has been an important topic of research. Titanium poses unique machining challenges, primarily due to its low thermal conductivity, causing high temperatures in a process zone during cutting, and high chemical affinity to tool materials, which can lead to welding of Ti particles to tools, accelerating adhesion-based tool wear. Several studies demonstrated that the β phase of Ti is particularly challenging to machine; in fact, the material's heat treatment condition significantly affects the overall machinability [1].

Typically, this problem is addressed by two approaches: (1) imposing low cutting speeds (often less than 50 m/min) and feeds (typically, below 0.3 mm/rev); (2) employing cutting fluids or coolants. The use of low cutting speeds inevitably lead to an increase in machining costs, especially when many aircraft components require almost 90 % of a blank material to be removed to obtain a finished product.

A. Roy (✉) · V. V. Silberschmidt
Wolfson School of Mechanical and Manufacturing Engineering, Loughborough University, Loughborough, Leicestershire LE11 3TU, UK
e-mail: A.Roy3@lboro.ac.uk

V. V. Silberschmidt
e-mail: V.Silberschmidt@lboro.ac.uk

J. P. Davim (ed.), *Machining of Titanium Alloys*,
Materials Forming, Machining and Tribology, DOI: 10.1007/978-3-662-43902-9_6,
© Springer-Verlag Berlin Heidelberg 2014

Additionally, for deeper cuts (i.e. high depths-of-cut magnitudes), a further low-ering of cutting speeds is required. Thus, there is a clear need to increase material removal rates (MRRs) to improve machining economics. In recent years, machining costs involving the use of cutting fluids have increased substantially, primarily due to environmental concerns. The handling of cutting fluids as well as their eventual disposal must obey strict rules of environmental protection. As a result, the costs related to cutting fluids represent a large amount of the total machining costs, which may exceed the cost of cutting tools itself [2]. Conse-quently, elimination of cutting fluids, if possible, can be a significant economic incentive.

Thus, there is a need for alternate techniques in dry machining of Ti-alloys. This is a reason for a spate of research in hybrid approaches to machining, whereby two or more complementary cutting techniques or mechanisms are used simultaneously to achieve greater productivity or enhance product quality. Examples of such approaches are a combination of laser-assisted machining (LAM) and cryogenic cooling of the tool [3]; electrical discharge machining (EDM) in milling of Ti alloys [4] and combination of EDM with ultrasonic machining [5], to name a few. One promising hybrid machining technique, which is the subject of this chapter, is ultrasonically assisted machining (UAM) and its variants.

The working principle of UAM is based on subjecting a machining tool to two independent motions. The first is a driving motion, which shapes the work-piece in a conventional process. Next, high-frequency (ultrasonic) vibration of specific frequency and intensity in a defined direction is superimposed on the driving motion of the tool. Thus, UAM belongs to the general class of vibration machining; however, it exhibits unique characteristics, which will be the subject of discussion in the following sections.

In this chapter, two specific hybrid machining processes will be discussed. First, the use of ultrasonically assisted machining in turning of Ti-alloys will be described. This process is henceforth referred to as *ultrasonically assisted turning* (or UAT for short). Two case studies with be presented with two different Ti-alloys. Then, a variant of the UAT, in which hot machining is combined with it to yield further improvements, is discussed. This machining process is referred to as *hot ultrasonically assisted turning* (or HUAT for short).

6.2 Ultrasonically Assisted Machining

Vibration machining (VM) was developed in the fifties of the last century for machining of ceramics and other hard and brittle materials [6]. There was wide-spread interest during the 1980s, with increased applications of ceramic materials in the industry. VM experiments with steel, glass and brittle ceramics confirmed that life of diamond tools could be extended with improvements in surface finish when compared to conventional machining [7]. Here, we make an important distinction from a traditional understanding of ultrasonic or vibration machining processes. Typically, ultrasonic machining is understood to be an abrasive process, in which material removal is purely mechanical. The process equipment consists of a vibrational horn (the sonotrode), a tool part, an abrasive paste (typically, with boron carbide particles), and the working material. During machining, the frequency is adjusted, so that the tool-sonotrode system resonates at around 20 kHz (thus making it ultrasonic); as a result, the abrasive particles suspended in slurry are propelled at the work surface, causing rapid erosion. In this sense, the abrasive paste ultimately 'cuts' the work-piece.

The ultrasonically assisted machining (UAM) process, which is the subject of this chapter, is a new hybrid machining process, which differs from a traditional understanding of 'ultrasonic machining'. First, it is a dry machining process, where no coolants or an abrasive paste is needed (though, a system with flood cooling can be used, if required). Next, the vibrating cutting tool interacts with the work-piece directly and cuts the material using a micro-chipping process. Kinematically, the system is different from that of a conventional machining process, as the cutting tool translates as in conventional machining but with superimposed vibro-impacts, leading to improved cutting conditions as will be demonstrated here. The advantages of this method are not *a priory* obvious, because machine-tool vibration (chatter) has to be vigorously suppressed in most cases. Interestingly, when an externally controlled vibration is imposed on a cutting tool, significant improvements in surface finish, noise and tool-wear reduction are observed. Prior studies of vibration machining have shown that, to achieve the maximum possible benefit from the vibratory cutting process, the vibration system needs to be tuned to resonance. One major complicating matter is that this resonance often depends non-linearly on the machining/processing parameters and the load experienced during the cutting process. These non-linear effects need to be accounted for in a robust design of control systems. A system with only frequency control is insufficient in achieving peak performance of an ultrasonic system and there is a need for autoresonant control systems in which the signal obtained from the performance sensor is fed to the transducer directly by means of a positive feedback which provides instant control of the mechatronic system [8].

In UAM several experimental studies have been carried out. The vast majority of reported results in UAM of Ti-alloys deals with the study of machinability of Ti-15-3-3-3, Ti-6246 and variants [9, 10].

6.2.1 Experimental Setup

Experiments were carried out on a universal lathe adequately modified to accommodate an ultrasonic cutting head with flexibility of switching between conventional and ultrasonic cutting regimes during a single turning operation. It is to be noted that such a cutting head may also be installed on a CNC machine. In the recent past, commercial variants have also been launched by DMG-Mori Seiki.

In the setup used to perform all the relevant experiments, the cutting head was a standard Langevin-type piezoelectric ultrasonic transducer mounted on a wave-guide with an aluminium concentrator, which amplifies the ultrasonic vibrations. A schematic of the ultrasonic cutting assembly is shown in Fig. 6.1. The assembly was fixed to the cross slide of the lathe by a specially designed tool-post attach-ment (Fig. 6.2). To record cutting forces during the machining process a dyna-mometer is attached to the cutting head. For analysis, the cutting force can be resolved into three orthonormal components in the x, y and z directions corre-sponding to the tangential, radial and feed directions, respectively (Fig. 6.2). Here, the tangential force is traditionally referred to as the primary cutting force.

For all machining trials, a cemented-carbide turning tools (SECO: DNMG 150608 MF1 CP500) with a nose radius of 0.8 mm with a low-depth-of-cut/ finishing chip breaker optimized for low feed rates was used. The tool material has a tough micro-grain structure suitable for intermittent cutting. The tool was mounted orthogonal to the work-piece axis so that the effective rake angle was approximately 14° and a clearance angle 0°.

6.2.2 Case Study 1: Ti-15-3-3-3

6.2.2.1 Work-Piece Material

The work-piece material, designated as Ti-15-3-3-3, belongs to a group of meta-stable β-Ti alloys, demonstrating significant precipitation-hardening characteris-tics. The alloy was solution-treated and aged by annealing at 790 °C for 30 min followed by air cooling, resulting in a β-phase state [9]. The Ti alloy was man-ufactured at the GfE-Metalle und Materialien GmbH in Nuremberg, Germany, as were all the other alloys described in case studies below.

6.2.2.2 Experimental Studies and Machining Results

Short experimental runs were conducted in our machining tests, in which con-ventional turning (CT) was immediately followed by UAT. Each experimental run was repeated 6 times to obtain reasonable statistics for our experimental data.

During vibration-assisted cutting it is important to monitor the cutting kine-matics. To this end, a laser vibrometer was used to monitor cutting-tool vibration during the experiment. The cutting head demonstrated spurious vibrations in radial and axial directions with amplitudes of ~ 1 and ~ 0.3 μm, respectively. This shows the difficulty to achieve a pure one-dimensional vibration system in transducer design and manufacture. Still, the vibratory amplitude in the primary cutting direction (tangential direction) was observed to be 10 μm for all the cutting depths, significantly dominating the overall cutting process. The cutting parame-ters used in the tests are listed in Table 6.1.

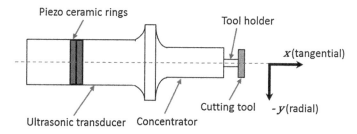

Fig. 6.1 Schematic of ultrasonic cutting assembly (see also Fig. 6.2)

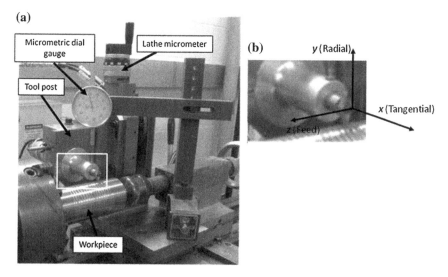

Fig. 6.2 **a** Ultrasonic cutting assembly; **b** zoomed-in picture of cutting tool (marked with white box in (**a**)) showing axis alignment

Table 6.1 Cutting parameters in experiments with Ti-15-3-3-3		
Cutting speed, V (m/min)		10–70
Feed, f (mm/rev)		0.1
Depth of cut, a_p (μm)		50–500
Vibration frequency, f (kHz)		17.9
Tangential vibration amplitude, a (μm)		10
Coolant		None

Superimposing ultrasonic vibration on the cutting tool is known to improve the surface finish of both ductile and brittle materials, with a concomitant reduction in cutting forces and machine chatter. It should be noted that imposing tangential vibration (Fig. 6.2) on the cutting tool in UAT changes the nature of the tool-work-piece interaction to an intermittent dynamic contact. From a 1-D analysis of such interaction, a relation between the critical oscillatory speed of the tool (v_c) and the speed of the work-piece motion (V) can be derived for the UAT process to be effective:

$$v_c > V. \tag{6.1}$$

The critical tool speed and the cutting speed are related to the machining parameters [8] by

$$v_c = 2\pi af \tag{6.2}$$

$$V = \pi n D \tag{6.3}$$

where n is the rotational speed of the lathe, D is the diameter of the machined work-piece, and a and f are the amplitude and frequency of the imposed vibration, respectively. For the machining parameters used, $v_c = 67$ m/min. It is expected that condition $v_c > V$ will ensure tool separation from the work-piece in each vibratory cycle.

6.2.2.3 Results: Cutting Forces

Cutting forces imposed on the tool were measured for CT and UAT performed with a varying depth of cut (a_p). The magnitudes of depths of cuts ranging from 50 to 500 µm were set with varying increments of 50 and 100 µm. A relatively low feed rate of 0.1 mm/rev was set to emulate high-precision machining, which typically deals with low material-removal rates (MRR) and, consequently, low feed rates. The raw data acquired with the dynamometer via an attached picoscope, was processed in Matlab, without any filtering, to obtain average cutting forces. In the analysis, data from the initial engagement was eliminated (see Fig. 6.3).

In all the experiments conducted, the axial/feed force component was measurably smaller than the primary tangential cutting force (Fig. 6.3 shows a typical force measurement from our experimental procedure). This is due to the low imposed feed rate coupled with a large nose radius operating at low depth of cuts. The measured cutting force components at different levels of a_p for both CT and UAT are presented in Fig. 6.4. The axial force component did not change significantly (Fig. 6.3) and thus was not used in the further analysis. The data in Fig. 6.4 presents average values obtained from multiple machining runs, and the error bars indicate the standard deviation of the measured forces. Significant force reductions (typically 75 %) in the primary and radial cutting direction in UAT are observed for the entire studied range of a_p. Improvements in the tangential cutting forces are expected in vibration-assisted turning as these correspond to the primary direction (along the x-axis) of ultrasonic vibration imposed on the cutting tool. Interestingly, high-frequency, low-amplitude vibration in the radial direction had

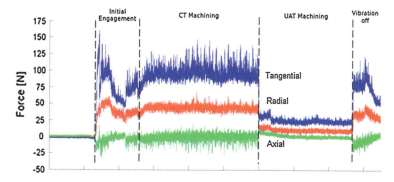

Fig. 6.3 Evolution of force-component signals produced by dynamometer in single run. Machining parameters used: $V = 10$ m/min, $a_p = 300$ μm, $f = 0.1$ mm/rev. In UAT: $f = 17.9$ kHz, $a = 10$ μm

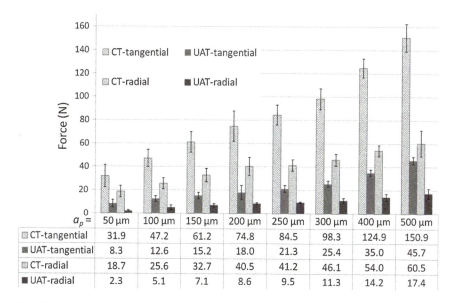

$a_p =$	50 μm	100 μm	150 μm	200 μm	250 μm	300 μm	400 μm	500 μm
CT-tangential	31.9	47.2	61.2	74.8	84.5	98.3	124.9	150.9
UAT-tangential	8.3	12.6	15.2	18.0	21.3	25.4	35.0	45.7
CT-radial	18.7	25.6	32.7	40.5	41.2	46.1	54.0	60.5
UAT-radial	2.3	5.1	7.1	8.6	9.5	11.3	14.2	17.4

Fig. 6.4 Cutting forces for CT and UAT at various depths of cut

also a noticeable effect on the measured cutting-force components. This is due to geometric reasons, since the curvature of the work-piece allows for tool separation in the radial direction, especially for small a_p. It is interesting to note that the average cutting forces in UAT for $a_p = 500$ μm are comparable to those in CT for $a_p = 150$ μm (see Fig. 6.4). This implies that, tool wear and tool life remaining the same in UAT, the MRR during vibration-assisted machining can be potentially increased by a factor >3 (owing to the diamond-shaped cutting tool geometry), with the cutting tool being exposed to the same level of cutting forces for these cutting depths. This hypothesis needs to be analysed in detail in the future.

Next, the effect of cutting speed on machining thrust forces was investigated. Figure 6.5 shows the measured forces in CT and UAT averaged over 5 experimental runs for each value of the used cutting speed. The cutting forces in CT show low sensitivity to the cutting speed within the studied range, as expected. However, in UAT, cutting forces increased with the increasing cutting speed, indicating that the derived relation (6.1) holds. In other words, with increasing speed, the tool separation in each vibro-impact reduces with a complete loss of separation at speeds exceeding v_c.

6.2.2.4 Results: Surface Topography

Characterisation of surface topography of the finished work-piece are presented for $a_p = 200$ μm and $V = 10$ m/min. Figure 6.6 compares the texture of typical surfaces machined with CT and UAT (presented as 2D field plots). Distinct periodicity can be observed for the conventionally turned surface whereas for the enhanced machining technique this regularity is somewhat curtailed. In CT, the direction of tool path during machining is evident, with a periodicity of some 100 μm, corresponding to the used feed rate of 0.1 mm/rev (Table 6.1). The machined surface profiles were analysed employing various texture parameters. Amplitude parameters calculated from the roughness profile, such as Ra, show a reduction of 49 % in UAT when compared to CT. This implies that within the measured sampling lengths the average roughness is significantly lower in UAT: the multiple cycles of reversed tool motion in UAT have a polishing effect on the machined work-piece surface.

6.2.2.5 Results: Sub-Surface Analysis

Conventional machining leads to high temperatures in the process zone and at the tool-work-piece interface. Coupled with low thermal conductivity of the β-Ti alloy under study, it was imperative to check a sub-surface layer of the work-piece for phase transformations. Usually, high temperatures in β-Ti alloys lead to formation of α-Ti phases that appear as needle-like structures under microscope. These phases are particularly undesirable as they compromise the improved mechanical characteristics of the β phase.

Sub-surface layers of work-pieces obtained with UAT and CT for machining conditions corresponding to $a_p = 500$ μm were analysed. Figure 6.7 compares the sub-surface microstructures for these two techniques. The alloy presents a coarse-grain structure with grains averaging 100 μm in size. The images show no needle-like (α-Ti) features and no visible changes in the grain size for both UAT and CT when compared to a virgin work-piece sample (i.e. prior to machining (Fig. 6.7a)). Since no visible changes were observed in the UAT work-piece (Fig. 6.7), it is safe to claim that no phase transformations are expected at the studied depths of cuts and cutting speeds.

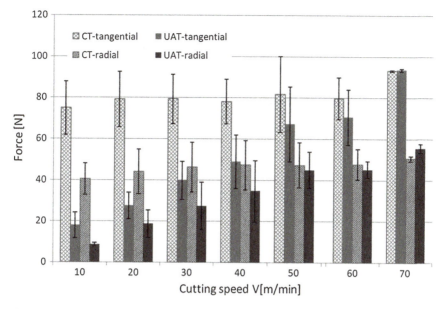

Fig. 6.5 Cutting forces for $a_p = 200$ μm in CT and UAT at various cutting speeds

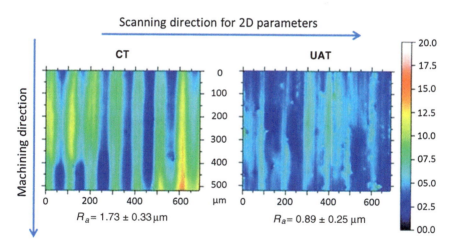

Fig. 6.6 Interferometry scan on area of 0.53 mm × 0.7 mm of surfaces machined with CT and UAT for $a_p = 200$ μm and $V = 10$ m/min

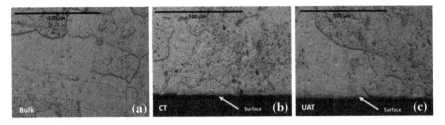

Fig. 6.7 Etched cross sections of work-pieces: **a** virgin-state bulk sample; **b** machined with CT for $a_p = 500$ μm; **c** machined with UAT for $a_p = 500$ μm

Table 6.2 Cutting conditions used in experiments with Ti-6246	Parameters	Magnitude
	Cutting speed, V (m/min)	10; 30; 60
	Depth of cut, a_p (μm)	200
	Feed rate, f_r (mm/rev)	0.1
	Vibration frequency in UAT, f (kHz)	20
	Vibration amplitude in UAT, a (μm)	10

6.2.3 Case Study 2: Ti-6-2-4-6

6.2.3.1 Work-Piece Material

The second work-piece material, Ti-6246, was also produced by the GfE-Metalle und Materialien GmbH in Nuremberg, Germany. After 2x vacuum arc re-melting, the alloy was forged in the two-phase field followed by air cooling, stress-relief annealing and stripping. Next, the alloy was re-melted once in a plasma-beam cold hearth melter followed by casting and stress-relief annealing. Microstructure studies revealed that the average grain size in the alloy was 147 ± 13 μm [10].

6.2.3.2 Experimental Studies and Machining Results

Short machining runs were conducted in the experimental study. The machining parameters of these runs are listed in Table 6.2.

6.2.3.3 Results: Cutting Forces

Application of ultrasonic vibration to the cutting tool brought a noticeable reduction in the cutting forces (Fig. 6.8). A significant reduction (74 %) in tangential component of the cutting forces was observed for application of vibration at the cutting speed of 10 m/min. There was also a considerable reduction (59 %)

Fig. 6.8 Cutting forces in CT and UAT of Ti-6246 at $a_p = 200$ µm for various cutting speeds

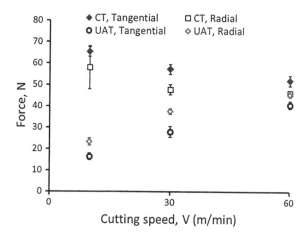

in the radial component of the cutting force at the same speed. The level of both cutting forces (the tangential and the radial) was slightly reduced with an increase in the cutting speed in CT in Ti-6246 (Fig. 6.8) (when compared to the cutting forces at 10 m/min). This effect was not visible in Ti-15-3-3-3. In UAT at $v = 30$ m/min, the reduction in the tangential and radial force components was observed to be 51 and 21 %, respectively. Consequently, at a higher cutting speed of 60 m/min, the reduction in tangential force component was 21 % with hardly any reduction in the radial force component. It should be noted that the critical velocity was ~ 75 m/min.

6.2.3.4 Results: Surface Topography

As before, UAT shows an improved surface quality in the machined work-piece (Fig. 6.9). The level of R_a for $V = 10$ m/min in CT was measured to be 0.82 ± 0.20 µm while for UAT it was 0.37 ± 0.04 µm. Thus, a significant improvement exceeding 50 % was observed in UAT when compared to CT. Similar reductions were also measured when machining at $V = 30$ m/min.

6.3 Discussion

The two case studies presented demonstrate that UAT shows systemic improvements in machining the Ti-alloys in general, with a significant measureable reduction in machining forces with an improved surface topography of the machined work-piece. Studies also indicate that UAT does not lead to any noticeable detrimental effects such as grain growth and re-crystallisation in machined work-piece.

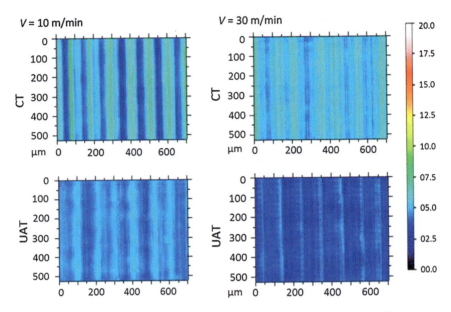

Fig. 6.9 Surface profile scan for CT and UAT at various cutting speeds at $a_p = 200$ μm

6.4 Hot Ultrasonically Assisted Machining

In this section, a new hybrid machining process is discussed that combines the advantages of UAT with the well-documented effects of hot machining in metals and alloys [11]. The experimental setup discussed in Sect. 6.2 was modified to house a band resistance heater around the cylindrical work-piece (Fig. 6.10).

6.4.1 Experimental Setup and Methodology

For hot-machining tests, a band-resistance heater, encapsulating the work-piece, was used as a heat source to increase the temperature of the work-piece to 300 ± 10 °C. Thermal measurements were performed using a Teflon-coated K-type thermocouple with a maximum measuring range of 1,200 °C in conjunction with a thermal camera (FLIR ThermaCAM^{TM} SC3000) in real-time.

The Ti alloy used was the β Ti alloy, Ti15-3-3-3, used in Case study 1; this allows for a realistic comparison with UAT.

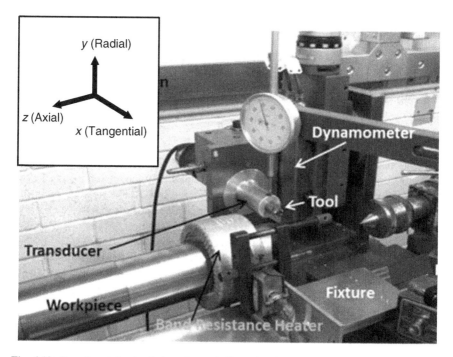

Fig. 6.10 Experimental setup in hot ultrasonically assisted turning

6.4.2 Experimental Studies and Machining Results

Hot machining was carried out at an elevated temperature of 300 °C, i.e. the work-piece was pre-heated before the machining process. Next, the band resistance heater was removed. The temperature at the work-piece surface was monitored continuously. Each experimental test lasted for approximately 90 s. Within the first 20 s the depth of cut was set to the desired magnitude followed by hot conventional turning (HCT) for 20 s. Next, ultrasonic vibration was switched on for approximately 40 s of the HUAT regime before being switched off to recover the HCT cutting conditions. Each experiment was repeated at least five times to obtain reasonable statistics on the experimental data. The machining parameters used in the tests are listed in Table 6.3.

6.4.3 Results: Measurements of Cutting Forces
and Temperature

Cutting forces were measured in real-time during the machining operation for various depths of cut. A substantial reduction in tangential and radial components of forces was observed in turning of Ti-15-3-3-3 using HUAT when compared to

Table 6.3 Cutting
parameters in HUAT

Cutting speed, V (m/min)	10
Feed, f (mm/rev)	0.1
Depth of cut, a_p (µm)	100–500
Vibration frequency, f (kHz)	20
Tangential vibration amplitude, a (µm)	8
In HCT and HUAT: Work-piece temperature, T (°C)	300, 500

conventional machining conditions, as reported for Case study 1 (Fig. 6.11). At
$a_p = 100$ µm, the reduction in tangential and radial components of cutting forces
was approximately 95 % in HUAT when compared to CT. The decline in cutting
forces reduces with an increase in the depth of cut and, ultimately, a uniform
reduction of 80–85 % was observed in HUAT $a_p > 200$ µm. On the other hand, a
reduction of some 20 % in cutting forces can be achieved in HCT. In hot
machining, the reduction in the cutting forces is mainly attributed to the decrease
in yield strength of the alloy at elevated temperature. However, in HUAT, thermal
softening when combined with tool separation in each vibratory cycle of tool
movement resulted in a significantly higher force reduction compared to that for
other machining processes.

Experiments were carried out to track the process-zone temperature in HCT and
HUAT. The temperature measurement at the process zone in CT was also conducted
during Case study 1, the results will be reported here for comparison. The temperature of the cutting region in HCT (where the work-piece was heated to 300 °C
before machining) at $a_p = 300$ µm, was approximately 250 °C higher when compared to that in conventional turning, whereas in HUAT the temperature was some
300 °C higher (Fig. 6.12). The temperature increase at the process zone with time
was observed to be gradual after the initial engagement of the tool. In HUAT, a
higher temperature in the process zone was observed when compared to HCT. This
is attributed to the temperature rise due to energy dissipation from vibro-impacts
imposed on the work-piece via the cutting tool in ultrasonic machining.

6.4.4 Results: Surface Topography

In this part of our study, surface roughness of the machined work-piece was
analysed for CT, HCT and HUAT at $a_p = 300$ µm. A significant reduction in the
roughness parameter (R_a) was observed in HUAT and HCT, when compared to CT
(Fig. 6.13). In HUAT and HCT, an improvement in excess of 50 % was observed.
Figure 6.13 compares the texture of typical surfaces machined with different
techniques (presented as 2D field plots). Distinct periodicity can be observed for
the conventionally turned surface whereas for the enhanced machining techniques
this regularity is somewhat reduced. The surface quality in HCT and HUAT was
effectively the same in statistical terms.

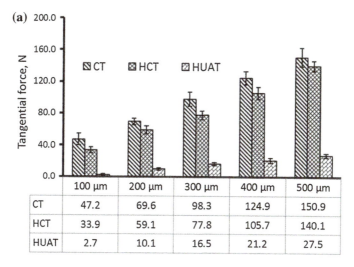

	100 μm	200 μm	300 μm	400 μm	500 μm
CT	47.2	69.6	98.3	124.9	150.9
HCT	33.9	59.1	77.8	105.7	140.1
HUAT	2.7	10.1	16.5	21.2	27.5

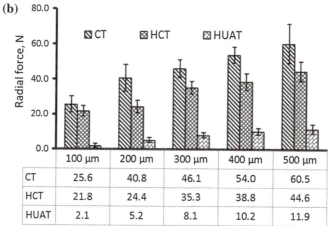

	100 μm	200 μm	300 μm	400 μm	500 μm
CT	25.6	40.8	46.1	54.0	60.5
HCT	21.8	24.4	35.3	38.8	44.6
HUAT	2.1	5.2	8.1	10.2	11.9

Fig. 6.11 Cutting forces at various depths of cut and $V = 10$ m/min; Hot machining at 300 °C. **a** Tangential force; **b** Radial force

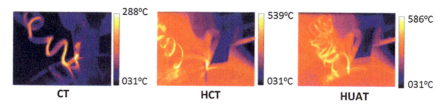

Fig. 6.12 Temperature of process zone in CT, HCT and HUAT

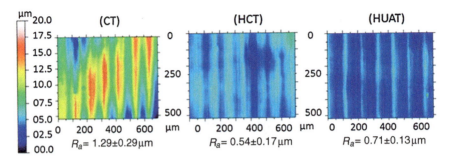

Fig. 6.13 Surface profile scan for different machining techniques at $V = 10$ m/min, $a_p = 300$ μm

6.5 Discussion

It is known that needle-like precipitate appear on the surface of Ti alloys when subjected to temperatures above 450 °C for more than 1 h. The specimens' machined with HUAT were investigated at different magnifications, and no signs of oxidation or metallurgical changes were observed. It should be noted that, though the temperature in the process zone was observed to be higher than 450 °C in HUAT and HCT, it did not show any detrimental effect on material's characteristics due to the short exposure time to high temperatures (in comparison to the mentioned 1 h). As a result, no precipitates were observed to form on the machined surface. The experiments indicate that it is possible to achieve cutting force reductions in excess of 80–85 %—when compared to conventional machining techniques—with an improved surface roughness of the machined work-piece material.

6.6 Conclusions and Outlook

Here, the advantages of using two specific hybrid machining techniques, employed to improve substantially machining of Ti-alloys, were discussed. The machining processes do not require any coolants, and have been shown to improve the topography of the machined surface with a significant reduction in machining forces. There is indeed a potential to increase MRR in ultrasonically assisted machining by several times.

However, there is a further need for research. First, advanced tools for ultrasonically assisted machining would benefit these techniques. The use of conventional tools in a vibro-impact machining process is not ideal; thus, tool manufactures need to investigate innovative tool geometry as well as tool materials and coating for UAM. Next, there is a need to understand the nature of the ultrasonic softening effect in alloys, if any. Finally, more research is required in developing next generation auto-resonant control systems, which will allow for widespread industrial applications and use of UAM.

References

1. Donachie MJ (2004) Titanium- a technical guide, 2nd edn. ASM International, New York
2. Klocke F, Eisenblätter G, Krieg T (2001) Machining: wear of tools, 2nd edn. Encyclopaedia of materials: science and technology. Elsevier, Oxford
3. Dandekar CR, Shin YC, Barnes J (2010) Machinability improvement of titanium alloy (Ti–6Al–4 V) via LAM and hybrid machining. Int J Mach Tools Manuf 50:174–182
4. Wang F, Liu Y, Zhang Y, Tang Z, Ji R, Zheng C (2014) Compound machining of titanium alloy by super high speed EDM milling and arc machining. J Mater Process Technol 214:531–538
5. Lin YC, Yan BH, Chang YS (2000) Machining characteristics of titanium alloy (Ti–6Al–4 V) using a combination process of EDM with USM. J Mater Process Technol 104:171–177
6. Isaev A, Anokhin V (1961) Ultrasonic vibration of a metal cutting tool. Vest Mashinos (in Russian)
7. Shamoto E, Moriwaki T (1999) Ultraprecision diamond cutting of hardened steel by applying elliptical vibration cutting. Ann CIRP 48:441–444
8. Astashev VK, Babitsky VI (2007) Ultrasonic processes and machines: dynamics control and applications. Springer, New York
9. Maurotto A, Muhammad R, Roy A, Silberschmidt VV (2013) Enhanced ultrasonically assisted turning of a β-titanium alloy. Ultrasonics 53:1242–1250
10. Muhammad R, Hussain MS, Maurotto A, Siemers C, Roy A, Silberschmidt VV (2014) Analysis of a free machining α + β titanium alloy using conventional and ultrasonically assisted turning. J Mater Process Technol 214:906–915
11. Muhammad R, Maurotto A, Demiral M, Roy A, Silberschmidt VV (2014) Thermally enhanced ultrasonically assisted machining of Ti alloy, CIRP J Manufact Sci Technol 7(2):159–167

Index

J. P. Davim (ed.), *Machining of Titanium Alloys*,
Materials Forming, Machining and Tribology, DOI: 10.1007/978-3-662-43902-9,
© Springer-Verlag Berlin Heidelberg 2014

CPSIA information can be obtained at www.ICGtesting.com
Printed in the USA
LVOW02*1916230814

400588LV00001B/24/P